African Epistemologies in Higher Education Research

I0885526

Bringing a needed perspective on African epistemologies on the critical topics of higher education in relation to knowledge systems, this book highlights how knowledge creation processes influence higher education systems, society, and African development.

This book uses an interdisciplinary approach to frame the connections between academic knowledge systems. Specifically, it seeks to answer questions on the trends in knowledge mobility, histories, and sociological dimensions in knowledge production in post-colonial Africa. The discussion explores how existing knowledge systems can better align with past and present narratives throughout African history and philosophies. The primary thought behind this book is to deconstruct the idea of a free market, the issue of corruption, racism, and the neoliberalist approach to knowledge creation and transmission. Thus, it seeks to answer questions on the history and sociological dimensions of knowledge production in higher education.

The book argues that African epistemologies can be better understood by investigating present sociologies and histories shaping African higher education research. Researchers and university students in the fields of sociology of education, economics of education, higher education, and policy will find this book very useful.

Kolawole Samuel Adeyemo is a nationally rated scientist by the National Research Foundation (NRF) of South Africa. He is an Associate Professor in the Department of Education Management and Policy Studies at the University of Pretoria, South Africa. He holds a PhD in Educational Management from Cavite State University in the Philippines and did a postdoc at the University of Pretoria. He is a researcher in the field of international and comparative studies in higher education. His work often involves exploring new populations of internationally mobile students, issues of policy in emerging and under-researched regional higher education destinations. He has worked as a visiting scholar at Indiana University Bloomington in the United States, Aarhus University in Denmark, and visiting professor at the University of Victoria in Canada. He is the South Africa Country Director for the Society of Transnational Academic Researchers (STAR) and Assistant Editor for the *Journal of International Students*, a top-ten ranked academic journal in higher education. His recent book, *Rhetoric of the Asia Higher Education Rankings* (2023), analyses the implications of the Quacquarelli Symonds (QS) world university rankings to gain insights into the usefulness of reputation rankings in addressing social inequality.

Routledge Research on Decoloniality and New Postcolonialisms
Series Editor: Mark Jackson, *Senior Lecturer in Postcolonial Geographies, School of Geographical Sciences, University of Bristol, UK.*

Routledge Research on Decoloniality and New Postcolonialisms is a forum for original, critical research into the histories, legacies, and life-worlds of modern colonialism, postcolonialism, and contemporary coloniality. It analyses efforts to decolonise dominant and damaging forms of thinking and practice, and identifies, from around the world, diverse perspectives that encourage living and flourishing differently. Once the purview of a postcolonial studies informed by the cultural turn's important focus on identity, language, text and representation, today's resurgent critiques of coloniality are also increasingly informed, across the humanities and social sciences, by a host of new influences and continuing insights for different futures: indigeneity, critical race theory, relational ecologies, critical semiotics, posthumanisms, ontology, affect, feminist standpoints, creative methodologies, post-development, critical pedagogies, intercultural activisms, place-based knowledges, and much else. The series welcomes a range of contributions from socially engaged intellectuals, theoretical scholars, empirical analysts, and critical practitioners whose work attends, and commits, to newly rigorous analyses of alternative proposals for understanding life and living well on our increasingly damaged earth.

This series is aimed at upper-level undergraduates, research students and academics, appealing to scholars from a range of academic fields including human geography, sociology, politics and broader interdisciplinary fields of social sciences, arts and humanities.

The Coloniality of Modern Taste
A Critique of Gastronomic Thought
Zilkia Janer

Coloniality and Decolonization in the Nordic Region
Edited by Adrián Groglopo and Julia Suárez-Krabbe

African Epistemologies in Higher Education Research
Kolawole Samuel Adeyemo

For more information about this series, please visit: https://www.routledge.com/Routledge-Research-on-Decoloniality-and-New-Postcolonialisms/book-series/RRNP

African Epistemologies in Higher Education Research

Kolawole Samuel Adeyemo

Routledge
Taylor & Francis Group
LONDON AND NEW YORK

First published 2024
by Routledge
4 Park Square, Milton Park, Abingdon, Oxon OX14 4RN

and by Routledge
605 Third Avenue, New York, NY 10158

Routledge is an imprint of the Taylor & Francis Group, an informa business

© 2024 Kolawole Samuel Adeyemo

The right of Kolawole Samuel Adeyemo to be identified as author of this work has been asserted in accordance with sections 77 and 78 of the Copyright, Designs and Patents Act 1988.

All rights reserved. No part of this book may be reprinted or reproduced or utilised in any form or by any electronic, mechanical, or other means, now known or hereafter invented, including photocopying and recording, or in any information storage or retrieval system, without permission in writing from the publishers.

Trademark notice: Product or corporate names may be trademarks or registered trademarks, and are used only for identification and explanation without intent to infringe.

British Library Cataloguing-in-Publication Data
A catalogue record for this book is available from the British Library

ISBN: 9781032428833 (hbk)
ISBN: 9781032428840 (pbk)
ISBN: 9781003364740 (ebk)

DOI: 10.4324/9781003364740

Typeset in Times New Roman
by KnowledgeWorks Global Ltd.

Dedication

This book is dedicated to God, the foundation of all knowledge and reason for my existence.

Contents

List of Tables

Acknowledgements

Firstly, I wish to acknowledge the National Research Foundation (NRF) of South Africa for the funding that assisted in completing this book project.

Secondly, I would like to express my appreciation to the University of Pretoria for granting me a sabbatical leave with funding as Visiting Professor to the Centre for Studies in Religion and Society (CSRS) at the University of Victoria in Canada where I managed to write most of the chapters in this book. Likewise, a special thanks to CSRS for hosting and shaping my ideas about focus of this book.

Thirdly, special appreciate goes to Professor Chika Sehoole, Dean Education Faculty at the University of Pretoria, and Professor Siphiwe Mthiyane, HOD: Department of Education Management and Policy Studies at the University of Pretoria, for their support throughout the period of this book project.

Fourthly, I want to appreciate Mrs Ophelie Dangbegnon, a PhD student at the Department of Education Management and Policy Studies at University of Pretoria, for working under very tight deadlines to make sure that the whole manuscript is well language edited.

Lastly, a very special appreciation goes to my wife, Mrs Mary Margaret Adeyemo, and son, Emmanuel Ife Adeyemo, for their continued supports in terms of prayers and understanding.

Introduction: An introduction to debates on African epistemologies

The theory of knowledge and the issue of global transfer of knowledge are deeply problematic and multidimensional. This book uses an interdisciplinary approach to frame the connections between academic knowledge systems. Specifically, it seeks to answer questions on the trends in knowledge mobility, histories, and sociological dimensions in knowledge production in post-colonial Africa. Understanding the trends and dimensions of knowledge produced by African academics without understanding the complexity of social stratification and related philosophical background can be challenging. Thus, the field of philosophy and sociology is employed to understand the dynamics of epistemic injustice in the higher education systems of Africa. It defines epistemic injustice broadly but frames the argument tightly around the problems of racism, corruption, capitalism, or neoliberalism. The assumption is that the construct of knowledge can be well studied and interpreted within the knowledge domains of social sciences and humanities.

The history of colonisation provides an exciting context for framing power and politics in Africa's globalisation knowledge and knowledge production. The legacies of colonists largely influenced the purpose and nature of knowledge in African higher education. Post-colonial researchers have argued the need to create an identity reflecting the uniqueness of African higher education systems. This includes rethinking colonial curricula, research publication, and knowledge usage for African development purposes. However, the history of knowledge in Africa cannot be erased but can be redirected to create independence in global knowledge transfer. The book deconstructs the idea of knowledge from a historical and anthropological lens to make conclusions on the future and purpose of African higher education based on its ability to produce knowledge for its context and sustainability.

The purpose of higher education has been framed wrongly in the literature as serving the economy. On the one hand, the overall perspective in academic literature also resonates with capitalist politicians' mindset. On the other hand, several higher education researchers are now challenging the narrow definition and framing of higher education as an agent of labour production for economic development purposes only. Higher education institutions are places of intellectual development. Among many roles of higher education is knowledge creation, which is vital for personal and social development, economic growth, and politics. The book examines what knowledge remains most relevant to the mission of higher education

DOI: 10.4324/9781003364740-1

and modern society's purpose. While the discussion is critical of capitalist and managerial approaches to knowledge production in higher education, the analysis is limited to the science of knowledge in Africa.

Therefore, the transformation of knowledge should begin with an overview of the history exchange and sharing. Higher education research and writing have been dominated mainly by a few names in Africa and abroad. Interestingly, many great higher education African scholars have studied in developed countries. African academics are yet to build acceptable frameworks and theories for African research. Thus, the frame of mind and thinking are still driven by many Western theories in higher education. This similarly translates to postgraduate supervision and the kind of doctoral graduates produced by African academics. Pragmatism and postgraduate research should promote the development of theories by African academics to understand some of the complex topics and issues in higher education. The argument is that people with lived experiences and an intellectual understanding of the continent can better conceptualise these theories. While global and Western theories in higher education cannot be ignored, the book emphasises the need to transform epistemic factors in knowledge creation. African epistemologies are necessary to ensure freedom in knowledge creation, dissemination, and transfer in African higher education research.

One crucial question in this book is that why it is difficult for Africa to redress the significant issues of inequality and inequity in knowledge dissemination. The literature in this field is focused on how epistemic justice or injustice is perceived in a particular context and how colonialism is problematic in knowledge dissemination in an African higher education setting. On the other hand, this book contributes to this literature by exploring the interdisciplinary subjects of philosophy, sociology, and history to reframe arguments of epistemic injustice in relation to knowledge systems in Africa. Essentially, it provides theoretical evidence of inequity in submitting articles and/or books for publication. It evaluates several studies published by Africans in Africa and foreign journals. Research publication capacity is framed based on the resources available to researchers seeking access to publish. The main argument is that Africa and its higher education policies are yet to recognise the consequence of promoting epistemic injustice in research, which Western epistemologies dominate.

Thus, this introduction provides a general background of the globalisation of knowledge and its implications for African higher education. Chapter 1 offers in-depth perspectives on transformation of knowledge from history of Africa. Chapter 2 engaged an existentialist epistemology literature to reframe purpose of higher education and the science of knowledge in Africa. Chapter 3 discusses the essence of applying pragmatism epistemology in different fields of study. Chapter 4 is critical of capitalist and managerial approaches to knowledge production in higher education. Chapter 5 concentrates on racism in higher education and approaches it from the standpoint of epistemological realism. It makes the case that racism entails stifling criticism. Finally, Chapter 6 offers discussion on academics, students, and institutions to reframe the argument about corruption in African higher education.

1 Transforming knowledge from history

Introduction

There is always an attempt to conceal African truth history, including cultural sources or artefacts that explain it (Heliodorus, 2013). However, ignoring African culture and history is no longer possible due to the growth of literature on African knowledge (Falola, 2017). In most cases, the history of Africa is linked to the slave trade, which is also shown in the colonial names of freed Africans in 1808–1862 (Anderson et al., 2013). Spanish and Portuguese slave traders label dark skin Africans Negro (Rahman, 2012) and enslaved Africans were given the title of African American (Blyden, 2019). Though many African countries gained independence in the 1960s, they were not independent (Falola, 2022) in terms of knowledge. Historians, mainly non-Africans, were not explicit in their writing about what happened during colonisation. However, Eckert (2019) asks whether global history shapes African history. Between 1969 and 1971, efforts and meetings were recorded to draft and publish African history (Niane & Ki-Zerbo, 1998). Recent developments to uncover Africa's true history are essential to understanding how to link the past to the present. However, another question is whether history is enough to decolonise the mind of Africans who have been oriented into colonial ways of thinking about different issues. The argument is that understanding the true history of Africa is essential in transforming knowledge, promoting Africanisation in writings and curricula, and addressing difficult topics in higher education systems. The following section asks about the history of knowledge and argues that all knowledge originates from Africa and, thus, cannot be ignored. While knowledge transformation is necessary, ethnonationalism, where differences still affect Africans' perspectives, may complicate the transformation process.

The realities of ethnonationalism

Archaeological evidence shows that Egypt was one of the most brilliant ancient civilisations in the world (Rose & Allen, 2018). Thus, all knowledge remains incomplete without understanding it from Africa. Yet, Africa seems to ignore its power in shaping and influencing the world through its history. Oral tradition shows that Africa has a long-established cultural institution and tradition

DOI: 10.4324/9781003364740-2

(Nkulu-N'Sengha, 2005). Whether not having access to accurate information affected how Africans think about knowledge and relate with each other remains to be discovered. Alternatively, confirming that Africans believe strongly in their history and see it as a power to change common problems affecting their societies is to be determined. For instance, Africa is unable to integrate their economics like Europe. Segregation is a severe obstacle as African countries wish to operate as a unit, but transforming knowledge through their higher education systems is equally problematic. The problem of transformation is due to the meaning of transformation in South Africa (Lange, 2020), inequalities and economic issues in some African countries (Dlamini, 2018; Molla & Cuthbert, 2018), legitimation of knowledge hierarchies (Ndofirepi, 2017), historical background (Gobodo-Madikizela, 2017), and post-independent issues (Arowosegbe, 2016; Mamdani, 2016; Mouton, Louw, & Strydom, 2012).

Pre-independent, the African past, and slave trade were bloody in terms of competition among enslavers and/or colonists. During this time, colonists controlled the population, economics, psychology and cultures, and relations between Africa, the South of the Sahara, and the Arab world (Harb, 2015; Thomas, 2012). Africans at that time desired to break from the oppression of the colonial power that made them powerless in knowledge and underdeveloped their communities. So, the passion for decolonisation or transformation pre-independent and is still active today. Today, Africans seem to think independence from colonial rules equals complete freedom. They later realise that colonists have created a system and culture that will make it almost impossible for Africa to unite in knowledge sharing. European imperialists created division among the ethnic groups in Africa (Johnson, 2002). For example, political conflicts in Ethiopia and DRC Congo, the issue of Boko Haram in Nigeria, ethnic division in South Africa, civil war in Cameroon, and ethnic wars in many other African countries can be attributed to the structure that colonists left behind. Due to the concerning nature of political systems in Africa, it is also difficult for higher education to develop. In this case, different ethnic groups are fighting for their interests in resource sharing and power.

Thus, the legacy of colonists in Africa and Africans' perspectives is the reason for the unresolved problems of poverty, inequality, corruption, racial discrimination, and others in Africa (Mbondenyi, 2009). Of most importance is higher education's ability to expose the issue to students and transform their minds to see Africa as one unit capable of redressing the legacy of the past. Such an endeavour will require sharing the knowledge of different colonisation experiences because different countries colonised Africans. This also plays an integral part in how African see each other or the reason for their inability to unite to transform knowledge. There have been some regional corporations and efforts to integrate Africa (Hartzenberg, 2011), but these are most effective as Anglophone, West African organisations, or SADC integration on the agenda. These organisations need to think beyond their space to firmly bring African knowledge and history into higher education to transform knowledge about their identity, values, and peace education. In other words, Africans need to share common knowledge about their histories because different opinions will continue to place Africans behind in global knowledge production.

However, different opinions about African histories are still a concern. The wrong methodological approach to studying African history is also a problem. For instance, some Africans did not see much reason to end their relationship with Europe during colonial rule. The era of free trade allowed new imperialism and rival capitalist monopolies that made Japan and Germany rival the United States in Africa (Harvey, 2017). Based on the history, the relationship between Africa and Europe underwent revolutionary changes. Africa faced a serious colonial challenge by 1880, affecting educated Africans (Boahen, 2020). Recently, different countries still want to recolonise Africans and African knowledge differently. Africa is now a battleground for various countries, including from Asia, to exploit trade and knowledge because Africans are yet to understand history and units to fight imperialism.

Africans were not only colonised because some African elites did not see colonisation as a problem since they were against African freedom, thus making the process of gaining freedom superficial. In other words, African elites reinforced Western cultures and education after independence. This is why Africans continue to study world history, not African history, English legal systems, and various subjects that mimic Africa pre-independence. So, the era of negotiation and treaty making between imperial power and African rules to recognise Africans as equal and honour the severity and independence of the African states and politics happened with hidden histories. What African elites signed with colonists before freedom did not seems to consider the diversity of Africa in promoting its independence in knowledge and advancement. In the recent century, social justice is popular in higher education research as a lens for understanding transformation and knowledge systems in Africa. However, Africa needs to look at the issue of social justice beyond a mere occurrence and trace its root cause to the behaviours of African elites with colonists before independence.

Social justice and discourse on knowledge systems

A series of historical events can help us understand the debate and discourse on knowledge systems. Since social justice and inclusivity issues are foregrounded as part of knowledge inequality, there is a need to flag the reason for African epistemologies in higher education research. In other words, this framing is vital to direct focus on one of the critical areas of knowledge production avenue in higher education. History should be woven into how different research topics in higher education are conducted and reported because African scientists can promote social justice in their writing, activism, and advocacy. Therefore, higher education and knowledge production scholarship can be framed within the larger human rights spectrum. However, distribution and proximity to power can be used to understand how social justice can be achieved through African scholarship.

In the present century, several problems in Africa and higher education systems find their solutions deeply rooted within political ideologies (Wolff, 2017).

One of the most critical concerns in this regard is the distribution of power and power relation in knowledge production in African higher education institutions. The twenty-first century seems marked by a worldwide drift towards a popular form of knowledge production that ignores Africanisation but promotes collaboration of the West into the knowledge domain systems of Africa. There is a need to reflect on the implications of proximity to the power of the West regarding the ability to still control knowledge in the post-colonial era. Likewise, we also need to reconsider how African knowledge can be subjected to validation from the West. This notion includes research in higher education and policy to determine the type of knowledge likely to be produced. Will higher education also respect a controversial research focus in terms of differences in Africa without requiring Western theories to analyse it? These questions can affect knowledge production in higher education but also speak to how history can continue to reflect the way we conduct research in higher education after independence. In this sense, advocating freedom for all in research is paramount. Such an idea can be linked to Plato's ethics of justice (Annas, 1999), in that each group in society is being educated in line with its talents to benefit the whole group. African higher education requires a transformation that allows its outcomes to benefit the entire continent, not a curriculum that serves certain ethnic groups or causes division and competition. Students have a set of different experiences, needs and goals. Institutions need to provide for differences that exist between individual students. The priority of higher education in a democratic African society should be to develop the individual to value their history and heritage.

For instance, Nazi education theory and practice in Germany were used as instruments by Hitler and his followers to mould a totalitarian society with only one leader, one party, and one people (Baldi, 2014; Roitberg, Baptista, & Gomes, 2020). This is not what Africa should do because knowledge transformation requires freedom to share ideas and collaborate without autocratic rules. In other words, there is a need for theory and practice that allows African academics and students to engage in knowledge about their continent with the view to developing it. Today, ethnonationalism in Africa in knowledge production looks like other forms of incipient totalitarianism in the ideological sense. African society tends to discriminate against other ethnic groups within Africa and the country based on their identities or political group. This can be seen with the resurgence of ethnonationalism. This issue manifests in different forms and becomes a primordial irrationalism that divides Africa regarding knowledge and innovative research. When made part of the higher education system, it leads to the indoctrination of dominant groups against others.

Africa is expected to approach knowledge production to redress the past's inequality, not expand it. The continent should help informally create a cultural identity through socialisation in a given ethno-cultural milieu and formally through discussions occurring during research. The next section speaks to the importance of African studies and indigenous knowledge systems in addressing the problem of corruption in Africa and African higher education. It starts by explaining knowledge and how we can prove African knowledge in higher education research domains. This includes promoting African epistemologies in science, social science, arts, and humanities subjects. It exposes theories underpinning arguments about African societal problems and how they can be addressed without necessarily ignoring science.

African studies and indigenous knowledge systems

Whether Africans can demonstrate their social and human sciences knowledge is still up for debate (Minga, 2021). For instance, cultural anthropology and geographic data about Africa provide a wealth of information about other academic disciplines. Meanwhile, forensic science has issues proving specific facts in court proceedings (Moss, 2021) without using science. What we know and how we know it has primarily been defined by various investigations and strategies to verify truths, regardless of the epistemological pathway (Kaplan, 1964). African researchers did not suggest that their knowledge was only based on superstitious beliefs (Emeagwali & Shizha, 2016). The content of African history and the method of creating indigenous knowledge were undervalued by the African worldviews and ideals shared by non-African writers. The history of Africa was initially given by western researchers, who had different meanings and goals from those of Africans. The intention was to denigrate and diminish African identity, education, culture, and language. Endeavours to think in African ways to discover the truth regarding African challenges have been decried as irrational, unscientific, and unjustifiable. A historical attempt to support claims that, for example, corruption is primarily an African sickness, as has been suggested by Western writers, may lead to more disputes. Also, Africans relied on traditional techniques that may not have followed standard science to communicate, transmit knowledge, and thrive before the advent of civilisation. The comment about the word 'civilisation' bring about different questions. For instance, did Africans need to be 'civilised'? Could African societies not be regarded as 'sophisticated' in their knowledge acquisition system in health and legal system? Finally, Africans expressed themselves, transferred knowledge, and survived based on indigenous methods that may not necessarily follow conventional science or civilisation narrative.

Therefore, there is a need to reflect and analyse the underpinning theories that inform knowledge in Africa broadly and beyond the exclusive thinking of the West or beyond science (Watts, 2017). In higher education research, several topics, such as corruption or racism, require history and knowledge of the past to understand the truth. These topics are controversial and, in most cases, may not be considered as truth unless history and southern theories are used to prove them (see Bam, Ntsebeza, & Zinn, 2018). In higher education, decolonisation is required in precolonial African histography (see Bam et al., 2018; Mngomezulu, 2004) to promote changes in research approaches in sciences, arts, humanities, and social sciences. The question about the theories of knowledge and decoloniality is important for problematic issues in higher education systems. In other words, the conception of reality within Africa and African society could be sustained through historical evidence. The trend of thoughts on these issues shows the urgency to appreciate an Afrocentric identity and methods in writing and sharing knowledge of critical topics and issues in African higher education.

The metaphysical nature of knowledge

The metaphysical nature of knowledge (Cassam, 2020) is the first reason for the argument about African epistemology. Metaphysics of knowledge seeks to answer metaphysical questions about knowledge and its place in the world as they arise in such epistemological pursuits (Kallestrup & Pritchard, 2017). Likewise, there is concern about whether African epistemology movements (Ndlovu-Gatsheni, 2018) can omit science or disregard it in explaining African realities. This problem underpins discourses in neo-colonial power and politics of knowledge. Scepticism has always been the first step to decolonising and creating new knowledge in higher education (Luckett & Blackie, 2022). African epistemology seems to be challenging science and removing the power of knowledge and its control from the West, motivating efforts to inhibit the expansion of knowledge from Africa. The argument is that what Africans claim to know as knowledge might just be an illusion if others cannot justify and believe it. The challenge is to determine how Africans would challenge Euro-Western epistemological dominance without losing axiological epistemic value (McDonnell, 2014) in African studies. Meanwhile, scientific logic requires these proofs to be logical, coherent, and pragmatic.

Specific African ways of knowing and solving issues previously required scientific inquiry. If African knowledge on different issues is perceived as independent of the knowers, how did the knowledge held by Africans during pre-colonial times become a reality? It should be highlighted that the knowledge acquired before colonisation was experience-based and could only be subjected to the reality of the indigenous people. One of the greatest challenges in African higher education systems is ensuring that its reality reflects the diversity of students, stakeholders, and agencies interested in higher education outcomes. Due to the diverging nature of both entities, student movements have called for the transformation and decolonisation of the African higher education curriculum. Meanwhile, there are also debates surrounding the feasibility of curriculum decolonisation in Africa without violating ideas in the globalisation or globalisation of knowledge. What is real and true regarding the curriculum in higher education in Africa needs to be justified in light of the African society values in the post-colonial era. If African self-development and innovation are the prevailing state of affairs, the decolonisation of knowledge should reflect a curriculum that recognises the continent's history and problems.

In the future, African epistemology from the theory of knowledge is essential in understanding the dynamics of issues such as corruption (Dibua, 2017; Masengwe & Dube, 2020; Muraya, 2020; Ndille, 2018). How can Africans justify or validate their knowledge in light of world views on the same subjects? Doing so requires the origins and methods related to the claims to be reported as experiences or proven by evidence. Knowing and sharing knowledge is critical in the higher education setting. It seems improbable to persuade either students or society that the generation and dissemination of knowledge should be based on experience rather than evidence. Using what can be observed in society as a starting point, African researchers have demonstrated the actuality of their issues in higher education. For instance, it is a fact that student fees are majorly the cause of riots at several African universities. This is also related to the widespread perception that African politicians do a poor job of managing the continent's economy, which has a negative impact on parents' ability to pay for higher tuition. On the other hand, it might be necessary to demonstrate that African politicians are generally corrupt, which could include evidence of homelessness, joblessness after graduation, and disparities in access to higher education based on gender. Though this may be the case for many Africans in African societies, providing evidence that these issues exist in higher education systems may require a pragmatic approach in which what we now know needs to be investigated utilising appropriate philosophical understanding on knowledge systems.

Post-colonial and philosophical views on knowledge systems

What do we mean, for instance, when academics discuss and write about inequality and the globalisation of knowledge (Galbraith & Choi, 2020)? More thought needs to be given to the implications of information sharing for Africa. This engagement will test African scientists' resilience in devising ways to validate African knowledge. This is significant because ideologies and histories are reshaping worldviews. However, philosophies of knowledge proofing (Magezi, 2018) are necessary for knowledge production and distribution. Different debates are being sparked by the globalisation of knowledge in Africa and beyond (Scholte, 2017).

Issue of student protests

African knowledge is fundamentally based on its philosophy (Masolo, 2019). African knowledge history begins with colonialism and leads to the resistance that resulted in freedom from the politics and power that oppress and control knowledge (Mudimbe, 2020). Prior to the modernisation of medicine, education, politics, and other fields, African culture and tradition were authenticated utilising traditional means. Older individuals displayed their wisdom in solving conflicts and passed on their knowledge to the next generation. Currently, a sizable body of African philosophical literature is theorising knowledge outside Marxism for the knowledge revolution (Zondi, 2021). These theories are considered anti-colonial and anti-globalisation because their premise lies in the partial or total disregard for science. In the case of Ethiopia, the argument challenges and is framed against US neo-imperialism during the late 1960s and early 1970s. Students at the time sought the destruction of colonial symbols to advance the revolution (Alemu, 2022). Student protests or political movements supporting the African theory of knowledge are unlikely to address the root causes of campus problem. The corruption in higher education are motivated by neo-liberal policies (see Badat, 2020; politics (Mac Donald, 2018), racism (Mogaji, Maringe, & Hinson, 2020), among others).

If Africa aims to address violent-related issues, student violence during protests reflects poor behaviour and may imply that Africans' solution to critical matters is violence. Riots among students are a common way to express their discontent with problems. In many situations, the riots take a political turn and highlight additional problems. Therefore, there are two sides to the debate over the morality of resorting to riots to remove neo-imperialist symbols from Ethiopia. First, while this approach might be legitimate for transformational goals, it is also problematic since deep-seated corruption means that students can only interact with the leadership and make their voices heard through rioting. Second, it demonstrates the necessity for an in-depth analysis of the issue and how Africans deal with it in this era of globalisation. When analysing discussions concerning the evolution of Africa pre-and post-colonialism, various perspectives from theories of knowledge are helpful. The main goal of these discussions should be understanding colonial symbols' significance in relation to student uprisings or protests, as well as determining the role of their removal in the advancement of society.

Proofing corruption

Higher education systems and their histories were designed to create an elite social class that would continue to dominate the economy and politics in the pre-colonial era (Woldegiorgis & Doevenspeck, 2013). In other words, the goal of higher education was to acquire riches or accumulate wealth. Africans' mindset is one of obtaining higher education credentials and achieving economic or political power to rule over others. Often, the majority is kept destitute to manipulate them. Wealth accumulation is a priority in pre-colonial and post-colonial higher systems, and in the absence of wealth, alternative ways are sometimes found to acquire it, including stealing (Molosi-France & Makoni, 2020). The legacy of colonialism is reflected in the premise on which higher education was developed in all colonised countries. This doesn't signify that Africans were born corrupt or that colonial higher education fostered this trait. However, the essentially wealth-accumulating nature of higher education puts Africans' minds in peril and reversing this trend is now proving difficult.

The proliferation of knowledge about corruption worldwide cannot be excluded from the various social movements aimed at granting African ways of knowing epistemic freedom. The cross-border movement of persons should have promoted an appreciation of varying viewpoints in higher education research. However, alternative solutions to other social issues have frequently been muted by Western knowledge, which still holds total hegemony. The various ideologies used to investigate epistemologies are under question. Regarding the concept of knowledge in Africa, the many viewpoints and perspectives lead to more debates than clarifications. For instance, there are disagreements about how Africa might demonstrate universal knowledge of their reality (Ndlovu, 2018). For instance, corruption is an abstract concept. Although corruption is not concrete, its effects can be seen in Africa in the form of poverty and sluggish progress, possibly due to the inefficient use of resources for higher education. Therefore, corruption is a human activity. That being the case, how can we demonstrate that African minds, and people are more susceptible to corruption and have influenced higher education's knowledge systems as a result?

Epistemic oppression is receiving attention in post-colonial literature as a sign of colonial authority in creating knowledge (Brunner, 2021), and as such, determining that colonial oppression had an impact on African thoughts and behaviour on resource management and corruption is a priority. Or is it possible to demonstrate that Africans lack resource management skills and are innately corrupt? Reviewing the implications of epistemic resistance movements (Bledin & Rawlins, 2016) for knowledge of more salient social science research challenges is imperative (Walker, 2020).

Concluding thoughts

It is important to assert that African philosophy transforms African history and knowledge. Thus, its impact needs to manifest through African knowledge and the promotion of African epistemologies. There is an argument that science has contributed to a more significant intellectual exchange of ideas than ever. While this is true, whether science recognises the origin of its epistemologies from the history of Africa is another question. In some instances, the foundation of science and scientific knowledge in different fields cannot ignore African history. In the early days, Africans initially discovered discoveries and innovations, though science gave those names and modernised them. Considering the dynamic social conditions in African nations today, African scholars are asking questions through their philosophies and trying to reimagine science from their perspectives. In higher education, placing the foundation of knowledge in the history of Africa is fundamental. Previously, many Western missionaries have used coercion to change how Africans think and create advanced knowledge. In this case, they introduced higher education systems where the progressive and intellectual place could drive their form of knowledge. Meanwhile, the objective was to undermine African knowledge and exclude early Africans from participating in advanced subjects that would allow them to embrace their indemnity and culture. This is what African epistemologies and philosophies are challenging. The idea is that knowledge involves principles and that principles must be weighted in conflict situations to determine an overriding principle.

References

Alemu, A. (2022). Demystifying the image: Anti-colonial concepts of the Ethiopian Revolution. *Comparative studies of South Asia, Africa and the Middle East*, *42*(2), 442–453.

Anderson, R., Borucki, A., Da Silva, D. D., Eltis, D., Lachance, P., Misevich, P. … Ojo, O. (2013). Using African names to identify the origins of captives in the transatlantic slave trade: Crowd-sourcing and the registers of liberated Africans, 1808–1862. *History in Africa*, *40*(1), 165–191.

Annas, J. (1999). *Platonic ethics, old and new*. New York: Cornell University Press.

Arowosegbe, J. O. (2016). African Scholars, African studies and knowledge production on Africa. *Africa*, *86*(2), 324–338.

Badat, S. (2020). Reproduction, transformation and public South African higher education during and beyond Covid-19. *Transformation: Critical Perspectives on Southern Africa, 104*(1), 24–42.

Baldi, G. (2014). Education in Nazi Germany. *German Politics & Society, 32*(4), 75–78. https://www.jstor.org/stable/43917486

Bam, J., Ntsebeza, L., & Zinn, A. (Eds.). (2018). *Whose history counts: Decolonizing African pre-colonial historiography*. Stellenbosch, South Africa: African Sun Media.

Bledin, J., & Rawlins, K. (2016, October). Epistemic resistance moves. In *Semantics and Linguistic Theory* (Vol. 26, pp. 620–640). Proceedings of SALT. http://journals.linguisticsociety.org/proceedings/index.php/SALT/article/view/26.620

Blyden, N. A. I. (2019). *African Americans and Africa: A new history*. New Haven, CT: Yale University Press.

Boahen, A. A. (2020). *African perspectives on colonialism*. Baltimore, MD: JHU Press.

Brunner, C. (2021).Conceptualizing epistemic violence: an interdisciplinary assemblage for IR. International Politics Reviews, 9(1), 193–212.

Cassam, Q. (2020). The metaphysical foundations of vice epistemology. In *Vice epistemology* (pp. 37–52). London, England: Routledge.

Dibua, J. I. (2017). *Modernization and the crisis of development in Africa: The Nigerian experience*. Hampshire and Burlington, VT: Ashgate Publishing Limited.

Dlamini, R. (2018). Corporatization of universities deepens inequalities by ignoring social injustices and restricting access to higher education. *South African Journal of Higher Education, 32*(5), 54–65.

Eckert, A. (2019). Scenes from a marriage. African history and global history. *Comparativ, 29*(2), 36–51.

Emeagwali, G., & Shizha, E. (Eds.). (2016). *African indigenous knowledge and the sciences: Journeys into the past and present*. Leiden, The Netherlands: Sense.

Falola, T. (2017). *Africanizing knowledge: African studies across the disciplines*. New York, NY, United States: Routledge.

Falola, T. (2022). Africa and the world after the Second World War. In *The Palgrave handbook of Africa and the changing global order* (pp. 59–73). Cham, Switzerland: Palgrave Macmillan.

Galbraith, J. K., & Choi, J. (2020). Inequality under globalization: State of knowledge and implications for economics. In E. Webster, I. Valodia, & D. Francis (Eds.), *Inequality Studies from the Global South* (pp. 33–52). London, England: Routledge.

Gobodo-Madikizela, P. (2017). Response to commentaries. *Journal of the American Psychoanalytic Association, 65*(2), 251–263.

Harb, C. (2015). The Arab region: Cultures, values, and identities. In *Handbook of Arab American psychology* (pp. 23–38). New York: Routledge.

Hartzenberg, T. (2011). Regional integration in Africa. *Available at SSRN 1941742*.

Harvey, D. (2017). The 'new' imperialism: Accumulation by dispossession. In *Karl Marx* (pp. 213–237). Oxford, England: Routledge.

Heliodorus, A. E. R. (2013). The revival of ancient historical traditions in black America: The four revised. (Eds.) Yaacov Shavit. In *History in Black: African-Americans in Search of an Ancient Past* (Vol. 36), 36–53. New York: Routledge.

Johnson, R. (2002). *British imperialism*. Cham, Switzerland: Springer Nature.

Kallestrup, J., & Pritchard, D. (2017). J. Adam Carter, Emma C. Gordon, Benjamin Jarvis (Eds.). Epistemic supervenience, anti-individualism, and knowledge-first epistemology. In *Knowledge first: Approaches in epistemology and mind* (pp. 200–222). Oxford University Press, Oxford, United Kingdom

Kaplan, A. (1964). *The conduct of inquiry: Methodology for behavioral science*. Transaction Publishers for 1964 publication: New Jersey.

Lange, L. (2020). Transformation revisited: Twenty years of higher education policy in South Africa. In *Transforming universities in South Africa* (pp. 39–59). Leiden, The Netherlands: Brill.

Luckett, K., & Blackie, M. A. (2022). Beyond epistemology: The challenge of reconceptualizing knowledge in higher education. *Teaching in Higher Education, 27*(8), 1018–1026.

Mac Donald, H. (2018). *The diversity delusion: How race and gender pandering corrupt the university and undermine our culture*. New York: St. Martin's Press.

Magezi, V. (2018). Public pastoral care as nexus and opportunity for a transformed practical theology within decolonisation discourse in South African higher education. *In Die Skriflig, 52*(1), 1–10.

Mamdani, M. (2016). Between the public intellectual and the scholar: Decolonization and some post-independence initiatives in African higher education. *Inter-Asia Cultural Studies, 17*(1), 68–83.

Masengwe, G., & Dube, B. (2020). Postcolonial trends of decolonising the religious education curriculum in Zimbabwe and South Africa. *Multicultural Education, 6*(5), 127–134.

Masolo, D. A. (2019). *African philosophy in search of identity*. Edinburgh, Scotland: Edinburgh University Press.

Mbondenyi, M. K. (2009). The right to participate in the government of one's country: An analysis of article 13 of the African charter on human and Peoples' rights in the light of Kenya's 2007 political crisis. *African Human Rights Law Journal, 9*(1), 183–202.

McDonnell, J. (2014). CHAPTER FIVE: Challenging the euro-Western epistemological dominance of development through African cosmovision. *Counterpoints, 443*, 98–116.

Minga, K. J. (2021). African discourses on the Africanization and decolonization of social and human sciences. *Journal of Black Studies, 52*(1), 50–76.

Mngomezulu, B. R. (2004). *A political history of higher education in East Africa: The rise and fall of the university of East Africa, 1937–1970*. Houston, TX: Rice University ProQuest Dissertations Publishing, 3122509.

Mogaji, E., Maringe, F., & Hinson, R. E. (2020). Marketisation in higher education in Africa: New directions for a decolonising continent. In *Understanding the higher education market in Africa* (pp. 283–295). London, England: Routledge.

Molla, T., & Cuthbert, D. (2018). Reimagining Africa as a knowledge economy: Premises and promises of recent higher education development initiatives. *Journal of Asian and African Studies, 53*(2), 250–267.

Molosi-France, K., & Makoni, S. (2020). A partnership of un-equals: Global South–North research collaborations in higher education Institutions. *Modern Africa: Politics, History and Society, 8*(2), 9–24.

Moss, S. (2021). Knowledge and legal proof. Tamar Szabó Gendler, John Hawthorne, and Julianne Chung (Eds.). In *Oxford Studies in EpisteMology* (Vol. *7*) 176–213. Oxford University Press, Oxford, United Kingdom.

Mouton, N., Louw, G. P., & Strydom, G. L. (2012). A historical analysis of the post-apartheid dispensation education in South Africa (1994-2011). *International Business & Economics Research Journal (IBER), 11*(11), 1211–1222.

Mudimbe, V. Y. (2020). *The invention of Africa: Gnosis, philosophy, and the order of knowledge*. Morrisville, North Carolina, United States: Lulu Press, Inc.

Muraya, A. N. (2020). *Corruption in Kenya and communities of interpretation: A sacramental pneuma-ontological perspective*. St. Thomas University, FL: ProQuest Dissertations Publishing, 27995873.

Ndille, R.N. (2018). Educational transformation in post-independence Africa: A historical assessment of the Africanization Project. *Preprints.org, 2018*, 2018080062. Doi: 10.20944/preprints201808.0062.v1

Ndlovu, M. (2018). Coloniality of knowledge and the challenge of creating African futures. *Ufahamu: A Journal of African Studies*, *40*(2), 96–112.

Ndlovu-Gatsheni, S. J. (2018). *Epistemic freedom in Africa: Deprovincialization and decolonization*. New York: Routledge.

Ndofirepi, A. (2017). African universities on a global ranking scale: Legitimation of knowledge hierarchies? *South African Journal of Higher Education, 31*(1), 155–174.

Niane, D. T., & Ki-Zerbo, J. (Eds.). (1998). *UNESCO general history of Africa, vol. IV, abridged edition: Africa From the twelfth to the sixteenth century* (Vol. 4). Berkeley: University of California Press.

Rahman, J. (2012). African Epistemology Encyclopedia of Black Studies. Eds. Molefi Asante and Ama Mazama. Thousand Oaks, CA: Sage, pp. 39–44.

Rahman, J. (2012). The n-word: Its history and use in the African American community. *Journal of English Linguistics*, *40*(2), 137–171.

Roitberg, G. P., Baptista, F. M., & Gomes, L. R. (2020). Expositions of "degenerate" art and music in Nazi Germany: reflections on totalitarian aesthetics and education. *Revista Digital do Laboratório de Artes Visuais*.

Rose, D., & Allen, R. (2018). *Ancient civilizations of the world*. Essex, England: Ed-Tech Press.

Scholte, J. A. (2017). *Globalization: A critical introduction*. London, England: Bloomsbury Publishing.

Thomas, M. (2012). *Violence and colonial order: Police, workers and protest in the European colonial empires, 1918-1940*. Cambridge, London: Cambridge University Press.

Walker, M. (2020). Failures and possibilities of epistemic justice, with some implications for higher education. *Critical Studies in Education*, *61*(3), 263–278.

Watts, D. J. (2017). Should social science be more solution-oriented? *Nature Human Behaviour*, *1*(1), 1–5.

Woldegiorgis, E. T., & Doevenspeck, M. (2013). The changing role of higher education in Africa: "A historical reflection". *Higher Education Studies*, *3*(6), 35–45.

Wolff, H. E. (2017). Language ideologies and the politics of language in post-colonial Africa. *Stellenbosch Papers in Linguistics Plus*, *51*(1), 1–22.

Zondi, S. (2021). The constraints for Marxism and African revolutions: Breaking bread with Ayi Kwei Armah. Sabelo J. Ndlovu-Gatsheni, Morgan Ndlovu (Eds.). In *Marxism and decolonization in the 21st century* (pp. 171–185). London, England: Routledge.

2 Existentialist epistemology and the purpose of higher education

Introduction

Humans are central to the knowledge production process in higher education systems. Moreover, Bertolin (2018) claims that substantial positive externalities could facilitate economic growth. Creating a purpose for our lives is at the heart of existential philosophy. Issues often arise as higher education is expected to take on the responsibility of giving every student a purpose. However, it is implausible for students' aspirations to always align with higher education institutions' purpose. Hence, various higher education institutions, such as science, management, and technology universities, have been created to showcase their offerings and target market. Despite these efforts, institutions' goals might still not be in alignment with students' needs. In most cases, higher education attempts to demonstrate its ability to produce the most excellent managers, doctors, and engineers who can help to drive a sustainable economy. While this is important for national development, it is crucial to understand the benefits of higher education outcomes and whether they meet graduates' requirements. Considering the positive externalities framing, society, and graduates, receivers of knowledge, are both expected to enjoy the benefits of higher education. The structuring of higher education outcomes for various potential benefits is made possible by positive externalities.

The argument is that receiving a higher education may not always be to find employment. The advantages could also include developing an intellectual society and creating safe communities. Therefore, the public purpose of higher education (Cuellar, Bencomo Garcia, & Saichaie, 2022) can be framed beyond economic purposes to focus on intellectual contribution to society. The onus falls on higher education to take responsibility for serving multiple roles, including leading the upliftment of community members' lives through its products, new approaches, or finding ways to understand students and their needs. Humans, as thinking beings, are expected to demonstrate the capacity to assess the knowledge produced by higher education (Dunne, 2015) and determine how this knowledge will be employed. Yet, existentialist epistemology holds that humans are not just thinking beings but also have the ability to feel, which can make pupils rational or irrational when applying their knowledge (Mozaffari & Jahanian, 2016). In this context, students have the ability to reason, create, and take ownership of their knowledge

DOI: 10.4324/9781003364740-3

based on their perception of what society needs or ought. Africans consider that the goal of learning is to advance humanity, not merely to further one's knowledge. So, knowledge aims to advance and safeguard all species throughout the cosmos and foster human development (Nkulu-N'Sengha, 2005). Yet, pragmatists think that science can address most human issues (Prasad, 2021). This view contradicts existentialist ideas, which hold that people are emotional beings and might impact how students think about knowledge, for example, throughout their studies.

The complexity of higher education's purpose is partly due to the students' emotions and capacity for self-creation. Higher education institutions in Africa assume to understand how to attribute value to education and students' life aspirations. Programmes are therefore developed, and students are enrolled to learn the necessary material. Knowledge production occurs in this manner due to institutions' ability to decide the realities to reconstruct or create using various disciplines. As a result of their inability to perceive how theoretical knowledge might help them make a difference, students have attempted to question higher education institutions' power on curriculum (see McGregor & Park, 2019). Students aspire to contribute to society and not only figure as workers. This necessitates a curriculum and method of instruction that enable them to comprehend how society might operate differently. If institutions exist for students, students' thoughts on the future of the economy, politics, and other topics may influence institutional changes. Thus, the question remains: how can institutions break free from politics of knowledge (see Igwebuike, 2013), epistemic pluriversality (see Song, 2023), and promote Africanisation of epistemology in Africa (see Ndofirepi, 2014)?

The following section discusses various viewpoints on how students might affect higher education institution policies. Students are not the bearers of power since higher education policies and reforms are also used to distribute power. Because they believe they are the most significant component of higher education systems, students occasionally demand power. In addition, institutions may overemphasise the need to position students at the centre of all institutional matters. As a result, prioritising students over staff well-being and allowing them to manage institutional affairs is extremely troublesome. Hence, the neoliberal ambition to marketise higher education includes treating students as paying customers. In light of declining public funding in Africa, the issue is how to identify and analyse the purpose of higher education. Another consideration is whether students' increased influence does not compromise the knowledge created in higher education. Considering all these ponderings, the following section engages several ideas of existentialist epistemology over the meaning of higher education.

Students as customers of higher education

It is debatable whether or not students are customers of higher education (Guilbault, 2018). It seems to reason that students who pay for their education are essentially paying customers purchasing a specific experience. Should they receive the customer experience, market orientation and methodology will significantly impact knowledge transmission? Decisions made by students and factors influencing higher education policy will also affect this way of thinking. Students' choices, in some ways, express their feelings and emotions about their beliefs and the knowledge they hope to obtain while enrolled in higher education institutions. Though sound, this reasoning becomes problematic when higher education is structured based on students' emotions. In other words, it would be helpful to clarify the idea of prioritising students: what does prioritisation entail for the academic freedom of the teaching staff operating in various fields of knowledge? How should institutions of higher learning prioritise their students while providing unwavering quality? These are genuine questions that could assist in determining the scope of student demands and the degree to which institutions have control over the type of knowledge and its outcomes in the higher education sector. Structural concerns, including politics, ethnic diversity, religion, and many more, influence knowledge transfer in an African context. These elements occasionally influence judgements about higher education's existence and polarised knowledge. On the other hand, government regulatory organisations are expected to consider the overall goals of higher education carefully. Additionally, it appears that these organisations pay little attention to students' aspirations and focus more on the agendas of the political parties in power with regards to higher education and how their policies will affect the existence and significance of higher education during their time in office. Meanwhile, students no longer want to be disregarded.

The type of institution and the courses students enrol in are linked to their preference for creating a specific meaning from their experiences. Should higher education aim to fulfil each student's experience, would quality remain uncompromised? Student recruitment is governed by regulations shaped by the institution's goals. However, institutional regulations might prove prejudiced towards students of a particular religion, ethnic group, or who fit the institution's objective. UNESCO has identified goals for education (Bergman, Bergman, Fernandes, Grossrieder, & Schneider, 2018), including the higher education sector. These goals seem holistic as they cover demographics, financing, inclusion, quality, and employment. Hence, African higher education systems must be reformed to achieve these objectives. Due to issues with resources, quality, inequality, and the difficulty in continuing their operations, particularly for private higher education providers, there is a misalignment and poor success graduation rate in Africa's higher education sectors (Mohamedbhai, 2014). The ability of African institutions to meet students' demands amid these difficulties is the other issue. In East Africa, for instance, the institution requires that the curriculum helps students understand their existence and reality (Mngomezulu & Hadebe, 2018). Free discussion is needed because students were not significantly involved in the curriculum development, as shown by the East African engineering curriculum (Mngomezulu & Hadebe, 2018; Taajamaa, Järvi, Laato, & Holvitie, 2018). The contrary could make it challenging for students to understand any curriculum. This requires a redefinition of what knowledge means for Africa from the viewpoint of the society visualised by students.

In practice, African higher education and policy on admission seems to provide a rigid process for students to access higher qualifications and preferred courses. Several African higher education institutions make the case that they intend to choose the best students for the correct courses and ensure that the number of applicants does not exceed the capacity of the institutions. Alas, certain applicants are often excluded from the process. Institutions and government policy tend to, therefore, disregard students' objectives and the kind of meaning they hope to produce. For institutions to create meaning for society and support the government's ambition for development, admission criteria must reflect that objective. How can institutions identify students' aspirations and help them achieve them through institutional policies? Despite the challenge posed by this question, it is essential to have innovative ideas and stretch the purpose of higher education beyond a traditional, rigid curriculum biased towards institutions' hopes for students and society.

The teaching methodology could also vary to enable meaning development and formation for accepted students. The courses that lecturers teach have their own significance to them, and they are designed in a way that they believe should also have meaning for the students. In many African higher education institutions, incorporating concepts and their interpretation into teaching and learning is not an option. According to this line of reasoning, existentialists' approach to education is democratic (Mozaffari & Jahanian, 2016). It includes discussing ideas with students rather than merely teaching them course content and instilling a culture of compliance. When students are free to participate in curriculum development, they can consider how it relates to their life and purpose. In this situation, students can take ownership of their responses on assignments and in formal exams because the conveyed ideas are relatable and difficult to forget. On the other hand, lecturers' meaning creation includes using textbooks as a teaching tool and basing assessments only on concepts covered in class. This is the issue with conventional higher education systems, which consistently stifle the ideas of applicants who meet the admissions requirements. Admitted students would rather fight to uphold their goals than consistently dispute college policies and knowledge systems. Alternatively, students could abide by their institutions' meaning-creation methodology and help them propagate it after graduation. In summary, higher education should be framed in relation to the well-being of students (Sherman, 2020) and African identity in the twenty-first century (Masolo, 2019). Thus, there is a need for the repurposing of higher education in Africa to address knowledge discrimination (Lebakeng, 2022). Likewise, it is students as rational beings, for example, that are supposed to assign value to the meaning or purpose of higher education (Imafidon, 2018). In this case, students as people who can think and make choices can affect epistemology in higher education systems (Roelofse, 2017).

The next section foregrounds the interrelation of students and the purpose of higher education in terms of the meaning ascribed to knowledge. It focuses more on students' interpretation and influence on the curriculum. This is mainly dependent on institutions' definition of their students. The section outlines potential issues that could arise when policies are not challenged. In this way, institutions develop meaning for students and society based on the current state of the political environment in the nation where the institution is located. This pertains to the vice chancellor's appointment and other essential roles in developing meaning. The central thesis is that higher education may offer inclusive experiences that can eradicate inequity, making students actual participants of the institutional community towards change while minimising the need for outside interventions.

Students and knowledge creation in higher education

African classrooms present complex issues, especially in African online teaching and learning (Brown & Nthoi, 2022). The changes that have occurred, especially during the COVID-19 pandemic, have led to the transformation of knowledge production in higher education into blended and, in most cases, online learning. However, this is not the focus of the argument. The main claim is that, whether through online or conventional instruction, different philosophies can impact students' learning. For instance, when students consider themselves clients who need to be catered to, it may result in the students placing various demands on the institutions in this situation. In other words, students might challenge their institutions' authority by employing the market approach tactic. Hence, institutions relying more on student fees for their viability are more likely to cater to students than those that receive significant government funding. As a result, a student protesting against injustice and subpar education might harm an institution's reputation. Students' use of this power may open up crucial discussions about the most valuable knowledge or students' requirements. Depending on the vice-chancellor and executive, the practice of not listening to students in African higher education may be slowly changing. In essence, addressing inequity in knowledge sharing and acquisition can be done when students are viewed as significant members of the institutions.

Conceptually, inequality and inequity issues also affect student experiences in sub-Saharan Africa (Lebeau & Oanda, 2020). Inequality and inequity in student experiences and knowledge dissemination could be interpreted based on students' experiences on campus. Experience on campus includes an institution's unwillingness to receive feedback from students. Such institutions will not allow students to influence their teaching methods and assessments. Thus, pedagogy theory might be helpful to redress inequality and injustice in terms of knowledge creation and dissemination on campus. Adopting pedagogical theory to address inequalities (Burke, Crozier, & Misiaszek, 2016) could help in rethinking the purpose of higher education and learning in Africa. Making students aware of institutions' hope to integrate them into their community can help them understand the value of higher education (Quaye, Harper, & Pendakur, 2019). In other words, knowledge can be presented to students, and they can recognise concepts already active in their thoughts. Acknowledging means that institutions of higher learning permit students to understand and explore the concepts they have in their minds.

Higher education should have anticipated that, through intuition, introspection, and insights, students would search within their own minds for a duplicate of the absolute for academic performance (Filmer, Molina, & Stacy, 2015). Students may wish to learn in this manner, but they cannot express themselves beyond their participation in class discussions. It implies that academics must be able to derive meaning from student interaction as well as the curriculum. Many education philosophers call for practical solutions to educational problems, such as inclusivity in knowledge creation in a multicultural society and a diverse classroom situation (Doucette, Sanabria, Sheplak, & Aydin, 2021). However, the notion of 'practicality' about inclusivity is open to analytic inquiry.

The major thrust of the philosophy of education and the meaning of the activities in higher education systems is (Walker & Gleaves, 2016) construction and theorisation of meaning formation in higher education systems. In this sense, educational philosophy, in terms of sociocultural theories, otherwise known as the pedagogy theory of learning in higher education, follows in the footsteps of the Socratic view of philosophy regarding virtue and ethics in serving students (Mortari & Ubbiali, 2021). Therefore, higher education will need to explain how students as thinking beings can be helped to shape the purpose and meaning of higher education activities without necessarily imposing knowledge but encouraging students to pursue it. The purpose of human existence in Africa (Akinola, 2023) is important to understand the dynamic of epistemology in higher education. Scholars (Loureiro & Moretti, 2021; Noman, 2018; Reoch, 2019; Rettová, 2021) have researched existentialism in relation to what it means for Africa in post-colonial higher education. Likewise, existentialism has been studied in relation to feminism (Grant, 2016; Perez, Saavedra, Jones, & Abril-Gonzalez, 2022), teaching in higher education (Herzog, 2018), Ubuntu ethics and decolonisation of knowledge in South Africa (Naudé, 2019), acknowledge of indigenous students in knowledge system (Bracamontes, 2017), and based on intellectual history (El Shakry, 2021). In all of these, there is evidence for need to redress epistemic injustice in higher education. Many African institutions have shown interest in transforming their curriculum, but in reality, its implementation has remained a challenge.

Instead, curriculum planning in many African higher education institutions is often poor and superficial. In Africa and elsewhere, higher education should be able to answer the following question: what knowledge is of most worth, and to whom? This understanding could guide the way academics think about and prepare for classes. Since many African countries' curricula and reading materials are prescriptive, it is necessary to refocus the goal of higher education for each African rather than emphasise presumptions about what African students should learn to obtain employment after graduation. The focus should be on students' aspirations when developing evaluation criteria and throughout the teaching and learning process. In other words, teaching in higher education should place students at the centre while maintaining the quality of higher education offerings. This is why we should not allow higher education students, for example, to cram the definitions of concepts but rather provide them with training to formulate their meaning during class activities, discussions, and examinations. This may require pairing students into diverse groups where they can learn and benefit from individual differences rather than see them as a problem. Also, students ought to be able to infer meaning for themselves from the lessons they are taught and from their interactions with other students. Creating meaning entails the interpretation of the past, such as thinking back on previous issues, those of a historical or mythological nature, as well as understanding what higher education means. However, African modernism (see Lamola, 2017) may have also influenced curriculum delivery in higher education in Africa. Social justice in knowledge production in the African context should mean the curriculum should speak to the reality of African situations (see Tshaka, 2019) and reject epistemological anarchism (see Tan & Lim, 2021) that may be interfering with African ways of knowing.

Political influences

There are several problems whose solution intrinsically involves political ideologies in African higher education systems. Political ideologies mean that external forces like governments influence what is taught and how it is delivered to students. In most cases, the government announces the type of skills needed for economic development. These skills are expected to be converted into programmes and delivered into the labour market. In an institutional context, this implies compliance with government policy for the sustainability of institutions. Higher education is also faced with challenges related to students being treated as paying customers with expectations of excellent service delivery. Hence, institutions must comply with governments' requests and students' aspirations. The impact of this on education and learning is a significant unknown. Academics may need to involve students in their learning to contribute to their own experiences and create meaning to achieve their expectations, even though this may not always comply with current legislation.

Notwithstanding, there is a need for approaches that can contribute to a curriculum that excludes politics but influence the quality and individual aspiration positively. This thinking may show that students can play important roles in driving policies and politics of higher education governance. In this manner, institutions would be placing students at the centre of teaching, allowing them to interpret the curriculum based on their life experiences. Therefore, stimulating students' minds through teachings is a critical change often indirectly disallowed by higher education national policies. Though not self-evident, this approach could play a significant role in implementing higher education policies. Higher education institutions are not tasked with providing students with interpretations or meanings about contemporary situations. Higher education institutions exist to enrich and promote open learning. Rather than viewing learning or students as serving any political objective, the goal of higher education should be to recognise individual aspirations in the sociocultural theories of learning. As academics prepare learning materials, lesson plans, lecture videos, and assessment development, this philosophy is anticipated to serve as a mental framework. Since the state eventually assumes ethical as well as political sovereignty over the education of its citizens, it is necessary to continue advocating for students not to be taught solely by the state but ultimately for the benefit of the state.

The first step in creating an inclusive higher education in Africa (Tate & Bagguley, 2017) is to examine academics' ability to handle social and ethnic issues in the classroom. These political concerns are challenging to teach, especially to students with first-hand experience. Social challenges have put to the test higher education's capacity to react to the changes conveyed by students through their interaction in the classroom or social settings of the institution, among all other issues that need consideration in higher education. Idealist metaphysics and epistemology are supported by the sociocultural theory of learning (Yucel, 2018). Beyond the obvious political, cultural, social, educational, and economic implications, this concept could aid in redefining the aim of higher education. Conversely, higher education could be explained from the viewpoint of the 'ordinary' African impacted by its development and existence. Though this concept has promoted in-depth discussions of the subject among students, there are difficulties in engaging in such conversations. In academic debates in the African context, the problem of diversity, and in particular the concept of race, is crucial. Teaching should refrain from prescribing or imposing ideas on students but encourage them to become critical thinkers.

Africa faces an epistemological problem in addressing student voice issues in a heterogeneous setting without contributing to more forms of discrimination. What exactly does it mean when students in Africa are said to be failing and not making use of the educational opportunities at their disposal? Does it imply that laws restrict students, or should higher education have the final say in all intellectual pursuits conducted within its borders? With this line of thinking, ways in which the curriculum contributes to students' continued lack of education after graduation can be found. Students' understanding of their education and credentials in their culture presents a barrier. In other words, students are continuously determining their own meaning. The expense of higher education should have yielded results meeting students' expectations while enabling them to contribute positively to society. As the government uses financing to alter institutions' and students' objectives, this is a challenging problem. The politics surrounding higher education funding, which may come from public funds or student loans, is the first causal factor of this challenge. The commitment attached to the origin of funds is the second factor determining the allocation of said funds. Finally, since decisions are not made for them, and higher education outcomes are not for each student's specific aspirations, students are likely to be excluded from these discussions.

Power relations and higher education stakeholders

What should be accomplished through education in Africa is the most crucial query. Regardless of the context, higher education's purpose has been misconstrued by stakeholders. Understanding the power dynamics between the government, institutions, and students is necessary to determine who has to be appeased or disregarded. In this relationship, students often voice concerns or make demands through rioting and other forms of protest. Rioting indicates that students refuse to remain silent and want their voices heard in the curriculum and institutional management. The response from universities frequently reveals a political declaration of intent to inform students. In other words, students are forced into a position where they may only be notified of decisions rather than participate. Consequently, students have been driven to want more than simply information on issues such as racial discrimination and decolonisation of a curriculum leading to frequent protests. Students want to be taken seriously and fulfilled. As a result, there is a need to reimagine knowledge based on race (see Suffla & Seedat, 2021), gender especially based on the experience of women (see Lemberger-Truelove, 2016), decolonisation of mind from slave mentality (see Umeh, 2021), and African perspectives (see Gordon, 2020; Okeregbe, 2016).

Students play a crucial role in educational institutions and are the driving force behind academic activity. Institutions would be wrong to disregard or refuse to demonstrate a sincere intent to satisfy them. Depending on outside factors like the government, students may be fairly treated, informed, or overlooked. In this situation, institutions may prefer to satisfy the government by offering programmes and following rules, even if doing so will have a negative impact on students. Institutions should try to consider how to foster a sense of student-centeredness among their students. This can be accomplished by holding meetings, on occasion, to demonstrate their willingness to speak with students while also letting them know that they will still comply with government demands. What would happen if the pressure from students increases and the institution either yields to their demands or ignore them and reports back to the government? Depending on the institution's stance, any action may be considered non-compliance and affect the level of funding received. In such instances, institutions may wish to take the lead and assume a position of intellectual leadership regarding the goals of higher education. This is challenging because institutions can no longer follow prescribed methods and ways of educating students. The situation might also differ should governments adopt a student-centred approach to higher education governance. How will institutions manage the students' growing influence while continuing to provide outstanding services?

This explanation shows how power, politics, and policies can shape the meaning of higher education. In Africa, this is a challenge and can be attributed to various causal factors depending on the country. In some countries, challenges faced may be due to history, the type of government, geographical location, and other factors. Regardless of the factors affecting decision-makers on the meaning of higher education, synergy is vital among and between stakeholders to ensure that higher education systems are redefined for members of society. To achieve this synergy, grassroots policies that support stakeholders in shaping an ideal society after graduation are needed.

Conclusion

The chapter discusses specific indigenous-informed methodologies concerning the purpose of higher education. There has been much talk about the need for and the place of African epistemologies. What is lacking is working out concrete approaches from these philosophical arguments. It is time to stop the advocacy and address how the African worldview informs research practice. Moreover, the chapter engages the purpose of higher education to make a strong case for the production of knowledge and the need for African epistemologies. Essentially, the literature in this field provides an overarching aim for the chapter focusing on

'African knowledge', which is an insightful discussion. The notion of 'African knowledge' is questionable but also debatable. The debates on this topic was analysed to re-examine issues of student participation in curriculum and inequality in knowledge production.

The discussion on student learning in higher education highlighted the importance of formulating problems from sociocultural learning theories and literature rather than intuition or experience of higher education's economic purpose. The focus was on the function and application of literature from the viewpoint of students forming knowledge and the significance of higher education in Africa. Similarly, there is a requirement to justify knowledge derived from student experience for validation. More importantly, the knowledge of meaning formed by students about higher education activities enables institutions to demonstrate a valid, accurate reflection of teaching and learning. Taking a closer look at instances where students call for curriculum revision in terms of language contributing to subpar academic achievement, higher education's goals will not be achieved until students' understanding of the instructional language is transformed. Otherwise, higher education risks being disregarded or discounted as a factor in the academic success of the twentieth century. In this case, students have difficulty understanding and learning in a language that is significant to bad academic performance, dropout rates, and poor learning at many African colonial universities. With such concerns, higher education should consider what this means for their purpose and how to use this understanding to allow changes specific to the student's meaning.

Critical discussions relevant to students' histories seem to arouse unfavourable emotions or feelings as academic work to develop their independence and capacity for critical thought. These circumstances should alter higher education's views on student learning and experience. Otherwise, students may question the motivation to allow uncomfortable theories or debates in class. The learning environment should offer a chance to address this problem in a way that enables students to communicate their feelings. Higher education institutions will then be able to comprehend how outside forces view their presence or approach to carrying out their purpose. Instilling this culture of intellectual engagement in students immediately in the classroom would be best.

Epistemological diversity in the higher education system requires higher education to ensure comprehensive learning through different learning methods. The innovative approach involves using technology to teach students from diverse economic and social backgrounds. Using multimedia applications to support learning during a crisis reflects higher education's mission. On the hand, this may also generate some problems. For instance, showing students footage of how other nations deal with issues could lead them to acquire false perceptions or cause tension in their thinking. Due to historical events, some students are underprivileged and members of minorities. Thus, teaching may be personalised for each student. Finding strategies to demonstrate that the goal of higher education is inclusive requires identifying students whose minds have been corrupted due to their negative experiences. This approach encourages success and throughput while allowing for cultural integration and social cohesion among students.

Students cannot be assumed to have extensive self-knowledge. Exposing students to the wisdom in the instruction allows them to start understanding, sharing, and extending it back to their minds.

Many vital points have been raised in the literature about how student meaning can become higher education purpose in the recent century concerning how we teach and theories used to teach complex classroom topics. This point reflects current changes in African higher education systems. Instead, the literature provided general information that does not suggest how Africa should be reconciling students with its mission. The essence of higher education is not to serve only students or the economy but to contribute knowledge to societal issues such as health problems, global conflicts, and health, among others. What is lacking is a description of how the information regarding higher education in Africa's goal might enable students to address societal and global problems without offering prescriptive solutions. Finally, several parts of the literature overemphasised the role of higher education in stimulating the economy. Unsurprisingly, African societies still struggle to understand the reasons behind students' acquisition of degrees for the sake of employment rather than to address issues they have been grappling with since the colonial era due to the existence of information misrepresenting the purpose of higher education. To encourage students to voice their opinions on challenging subjects, institutions will need to foster an open culture. Critical problems concerning students should be examined and rethought in intellectual dialogues inspired by the instructional methodology. Research shows that a culture of open debate is not common or perhaps not encouraged in many African universities.

For this reason, students sometimes find it strange when asked to discuss issues such as race and religion. This kind of discussion should be encouraged in academic settings and the classroom. The purpose of such discussions is to improve students' capacity for thought and assist them in reconsidering the definition of higher education in Africa. Furthermore, students can use their experiences in higher education to support African development. Therefore, understanding the cultural perspective of the learning environment should be the primary priority of teaching and learning interactions. Higher education should redefine its meaning to speak to the cultural view of African society and the students they have admitted.

References

Akinola, F. A. (2023). An African appraisal of Sartre's existentialism and its implications for human existence in Africa. *OCHENDO: An African Journal of Innovative Studies*, *4*(1), 97–111.

Bergman, Z., Bergman, M. M., Fernandes, K., Grossrieder, D., & Schneider, L. (2018). The contribution of UNESCO chairs toward achieving the UN sustainable development goals. *Sustainability*, *10*(12), 4471.

Bertolin, J. (2018). Higher education and development in the knowledge society: From integral education to substantial positive externalities. *Higher Education for the Future*, *5*(2), 122–141.

Bracamontes, B. I. (2017). *A write to know: Using autoehnographic writing to explore marginalization and sense of belonging in community college students.* Fullerton: California State University.

Brown, B. A., & Nthoi, O. (2022). Cultural influences in online pedagogy and the final ends of African higher education. In *The emerald handbook of higher education in a post-Covid world: New approaches and technologies for teaching and learning* (pp. 289–314). Bingley, England: Emerald Publishing Limited.

Burke, P. J., Crozier, G., & Misiaszek, L. (2016). *Changing pedagogical spaces in higher education: Diversity, inequalities and misrecognition*. New York: Routledge.

Cuellar, M. G., Bencomo Garcia, A., & Saichaie, K. (2022). Reaffirming the public purposes of higher education: First-generation and continuing generation students' perspectives. *The Journal of Higher Education, 93*(2), 273–296.

Doucette, B., Sanabria, A., Sheplak, A., & Aydin, H. (2021). The perceptions of culturally diverse graduate students on multicultural education: Implication for inclusion and diversity awareness in higher education. *European Journal of Educational Research, 10*(3), 1259–1273.

Dunne, G. (2015). Beyond critical thinking to critical being: Criticality in higher education and life. *International Journal of Educational Research, 71*, 86–99.

El Shakry, O. (2021). Rethinking Arab intellectual history: Epistemology, historicism, secularism. *Modern Intellectual History, 18*(2), 547–572.

Filmer, D., Molina, E., & Stacy, B. (2015). What goes on inside the classroom in Africa? Assessing the relationship between what teachers know, what happened in the classroom, and student performance. *Paper for the Service Delivery Indicators Initiative*. Washington, DC: World Bank.

Gordon, L. R. (2020). Sartre's influence in black existentialism. In *The Sartrean mind* (pp. 501–514). London, England: Routledge.

Grant, J. (2016). Experience and epistemology in feminist theory. *The Oxford handbook of feminist theory*. New York, NY: Oxford University, 227–246.

Guilbault, M. (2018). Students as customers in higher education: The (controversial) debate needs to end. *Journal of Retailing and Consumer Services, 40*, 295–298.

Herzog, A. P. (2018). *An existentialist approach to teaching writing: Anguish, bad faith, and seriousness in composition*. Doctoral dissertation, University of South Carolina.

Igwebuike, T. (2013). Epistemic motivation for conceptual change in integrated science classrooms in non-Western cultures. *IISTE: Developing Country Studies, 3*(1), 43–50.

Imafidon, E. (2018). *African philosophy and the otherness of albinism: White skin, black race*. New York: Routledge.

Lamola, M. J. (2017). African postmodernism: Its moment, nature and content. *International Journal of African Renaissance Studies-Multi-, Inter-and Transdisciplinarity, 12*(2), 110–123.

Lebakeng, T. J. (2022). Repurposing the university in Africa against an explicit racist epistemology. *Journal of Humanities, 30*(2), 23–43.

Lebeau, Y., & Oanda, I. O. (2020). *Higher education expansion and social inequalities in sub-Saharan Africa: Conceptual and empirical perspectives* (No. 2020-10). UNRISD Working Paper.

Lemberger-Truelove, T. L. (2016). *Black women's search for meaning: An existential portraiture study on how Black women experience the 4 existential givens*. Doctoral dissertation, The University of New Mexico.

Loureiro, C. W., & Moretti, C. Z. (2021). Paulo Freire in abya Yala: Denunciations and announcements of a decolonial epistemology. *Praxis Educativa, 16*, 1–19.

Masolo, D. A. (2019). *African Philosophy in search of identity*. Edinburgh, England: Edinburgh University Press.

McGregor, R., & Park, M. S. A. (2019). Towards a deconstructed curriculum: Rethinking higher education in the global North. *Teaching in Higher Education, 24*(3), 332–345.

Mngomezulu, B. R., & Hadebe, S. (2018). What would the decolonization of a political science curriculum entail? Lessons to be learnt from the East African experience at the federal university of East Africa. *Politikon, 45*(1), 66–80.

Mohamedbhai, G. (2014). Massification in higher education institutions in Africa: Causes, consequences and responses. *International Journal of African Higher Education, 1*(1), 60–83.

Mortari, L., & Ubbiali, M. (2021). Service learning: A philosophy and practice to reframe higher education. *Athens Journal of Education, 8*(2), 115–138.

Mozaffari, M., & Jahanian, R. (2016). Identifying existentialist philosophy of education. *International Academic Journal of Social Sciences, 3*(8), 18–28.

Naudé, P. (2019). Decolonising knowledge: can Ubuntu ethics save us from coloniality? (Ex Africa semper aliquid novi?). *Decolonisation in Universities: The politics of knowledge*, 217–238.

Ndofirepi, A. P. (2014). Africanisation of epistemology in the 21st century university in Africa. Edward Shizha (Ed.). In *Remapping Africa in the Global Space: Propositions for Change* (Vol. 1, pp. 151–165).

Nkulu-N'Sengha, M. (2005). African epistemology. In M. Asante & A. Mazama (Eds.), *Encyclopedia of black studies* (p. 4045). Thousand Oaks, CA: SAGE Publications.

Noman, A. S. M. (2018). *Africological reconceptualization of the epistemological crises in postcolonial studies.* Doctoral dissertation, Temple University. Libraries.

Okeregbe, A. (2016). The African predicament and a case for Singer's 'Samaritanism': An existentialist interpretation. *Filosofia Theoretica, 5*(2), 19–36.

Perez, M. S., Saavedra, C. M., Jones, N. N., & Abril-Gonzalez, P. (2022). Woman of colour feminisms as a vehicle for racially-just methodologies in educational research. *International Journal of Research & Method in Education, 45*(4), 410–420.

Prasad, M. (2021). Pragmatism as problem solving. *Socius, 7*, 1–13.

Quaye, S. J., Harper, S. R., & Pendakur, S. L. (Eds.). (2019). *Student engagement in higher education: Theoretical perspectives and practical approaches for diverse populations.* New York: Routledge.

Reoch, A. (2019). An interview with the NEEM ("non-existent existentialist memes") admins. Alfie Bown, Dan Bristow (Eds.). In *Post-Memes. Seizing the memes of production* (pp. 233–251). Punctum Books.

Rettová, A. (2021). 9 Epistemology and literature: Positivism, indeterminacy, holism and relativism in the Swahili novel. Alena Rettová, Benedetta Lanfranchi, Miriam Pahl (Eds.). In *Critical conversations in African philosophy* (pp. 168–190). Routledge, London.

Roelofse, J. R. (2017). A paradigm shift in pre-theoretical deliberations on crime within spiritual existentialism. *Theologia Viatorum, 41*(1), 48–74.

Sherman, G. L. (2020). Existentialism and higher education: A renewed intersection in well-being. *Journal of Humanistic Psychology* (online first). 0(0). https://doi.org/10.1177/0022167820917231

Song, Y. (2023). "Does Chinese philosophy count as philosophy?": Decolonial awareness and practices in international English medium instruction programs. *Higher Education, 85*(2), 437–453.

Suffla, S., & Seedat, M. (2021). Africa's knowledge archives, Black consciousness and reimagining community psychology. *Decoloniality and Epistemic Justice in Contemporary Community Psychology*. Community Psychology. Springer, Cham. https://doi.org/10.1007/978-3-030-72220-3_2

V. Taajamaa, A. Järvi, S. Laato and J. Holvitie, "Co-creative engineering curriculum design — Case East Africa," 2018 IEEE Frontiers in Education Conference (FIE), San Jose, CA, USA, 2018, pp. 1–5, doi: 10.1109/FIE.2018.8658521

Tan, S. Z. K., & Lim, L. W. (2021). An existentialist approach to authentic science. *IBRO Neuroscience Reports*, *11*, 52–55.

Tate, S. A., & Bagguley, P. (2017). Building the anti-racist university: Next steps. *Race Ethnicity and Education*, *20*(3), 289–299.

Tshaka, R. S. (2019). The advocacy for Africanity as justice against epistemicide. *Black Theology*, *17*(2), 132–149.

Umeh, G. F. (2021). Communalist existentialism: A panacea for slave mentality in Africa. *Journal of African Studies and Sustainable Development*, *4*(3), 2–18.

Walker, C., & Gleaves, A. (2016). Constructing the caring higher education teacher: A theoretical framework. *Teaching and Teacher Education*, *54*, 65–76.

Yucel, R. (2018). Scientists' ontological and epistemological views about science from the perspective of critical realism. *Science & Education*, *27*(5), 407–433.

3 Epistemology of pragmatism in postgraduate research

Introduction

According to the pragmatic approach to knowledge production, solving a problem is not challenging as long as it can be identified (Prasad, 2021). Despite knowledge verification or validation and its use in solving complex issues, knowledge can still be ignored. Knowledge rejection often occurs because the offered solutions may not comply with the scientific inquiry process and methodology required to discover the truth. This mindset is the cause of the difficulties in altering postgraduate research in higher education, particularly doctoral education, which appears to follow similar knowledge production practises from the colonial era or firmly adheres to the outdated European model of supervision (Björkman, 2018; Rensfeldt, 2013). Postgraduate research supervisors often disagree during postgraduate research defences (Gullbekk & Byström, 2018) due to differences in problem-solving techniques and procedures, including research design. Producing knowledge requires the use of different methodologies depending on the type of research. Nevertheless, pragmatism holds that various practical approaches to discovering the truth exists, no matter the problem under discussion. The application of this ideology is complex because bureaucrats and some anti-intellectual policies in higher education heavily regulate the research process. What is the purpose of doing research and why there are diverse views regarding it higher education (Haapakoski, Hellsten, & Malone, 2019)? Researchers have tried to suggest epistemological frameworks and pragmatic ways of conducting research (Kelly & Cordeiro, 2020). There is a suggestion to focus on indigenous methods to solve African problems (Oppong, 2013). Other studies such as in South Africa (Davis-Warra, 2017), Uganda (Whyte & Whyte, 2015), and Tanzania (Vavrus & Barlett, 2012) argued for research processes that recognise diversity and promote inclusivity in higher education settings.

The following section focuses on postgraduate research in higher education, pragmatic epistemologies (see Dewey & Imagination, 2020), and indigenous knowledge. The argument is that there are several ways to create knowledge. Students can be crucial in expanding our thinking about different topics and improving our understanding of complex issues. How can students be supervised so that theories and methodologies will not be limited to conventional thinking

DOI: 10.4324/9781003364740-4

about conducting scientific research? This aspect is unravelled to make a case for southern theories, methodologies, and thinking in conducting research relevant to African situations.

Indigenous postgraduate research

Many scientific issues have non-scientific solutions in various fields. The pragmatic philosophy, meant to transform people's perspectives on creating new knowledge and information in multiple sectors, is effectively rejected by a political environment promoting rigidity of thought, which rejects answers to modern problems outside of science. Nonetheless, it is essential to assist postgraduate students in defining and identifying the concepts they are interested in pursuing during their postgraduate research supervision. These concepts may often be contrary to the supervisors' knowledge, experience, and habits. Therefore, reading is crucial for laying the groundwork for new information, but literature assertions should not be taken at face value. Students who use the programmatic approach must define and interpret several ideas from the literature (Kaushik & Walsh, 2019). The supervisor oversees this evaluation process, and their sole goal is to assist students in finding significance rather than impose their ways of thinking or study methods.

Demonstrating that African indigenous knowledge based on experience is a valid way of knowing is essential (see Alemu, 2018; Stilwell, 2010). The question remains to what extent postgraduate supervisors can prove to students the knowledge that cannot be defined is also knowledge. In the humanities, there are historical and religious topics in which knowledge foundation can only be described using indigenous languages and cues to better understanding. The method of knowing offers Africanisation and useful ways of comprehending African knowledge; it is neither universal nor based on science. Challenges arise when framing research issues entrenched in African culture or tradition using non-indigenous techniques. Various currents of thought call for popularising historical approaches to research indigenous matters. It is only possible to study issues such as marginalisation, violence against women, land rights, and all forms of maltreatment of indigenous people through the tradition and experience of those who have experienced marginalisation.

One approach to indigenous research is storytelling. Telling a story about past occurrences provides rich data about the incident and the state of mind of the victim in the present time. In this way, indigenous people who have been victims of rape, discrimination, oppression, and other forms of abuse can tell and retell their stories based on their observations and the change they would like to see going forward. In postgraduate research, storytelling is not very popular and sometimes not well adopted. Using such a method for indigenous research topics allows participants to share information about the events without intervening in the details. Second, questioning could influence how the participants may share the experience. In Africa, certain topics are culturally considered taboo. Sex and rape, for instance, are still not topics that many Africans wish to discuss, particularly for female participants interviewed by male researchers. Lastly, it is valid yet problematic to assume that telling tales of previous events can reveal the truth about indigenous people and their past. This is because it raises questions about who is equipped to create and implement the procedures for gathering such data. Therefore, Africans can highlight colonialism and uphold post-colonial independence through the use of storytelling as a method of inquiry.

The evaluation and analysis of actual photos or videos of situations related to indigenous people present an alternate method to demonstrate Africans' ability to provide accurate knowledge. The analysis from art galleries and/or actual cultural displays, such as ceremonies related to past experiences, is essential to uproot issues pertaining to African experience and knowledge. This type of research may be problematised outside science or conventional method of interviewing or conducting surveys to understand the people who have first-hand information about certain topics. The issue is how to eliminate bias in data collection that might be affected by emotion or a recent incident of abuse. Such data may not always be impartial or objective, bringing their dependability and applicability into question. Another concern is the researcher and whether or not their background may affect the narrative and data reporting.

Generating knowledge using the indigenous African method primarily depends on the researchers. Those with non-African/indigenous origins researching Africans and their challenges outside their society may not use African-recognised ways of arriving at the truth. If they employ African techniques, their emotions and judgement would need to be evaluated to ensure objectivity in reporting the facts. Researchers need to interpret the data by delivering stories or observing environments. Another technique is necessary when evaluating indigenous data to verify that the information is accurate and applicable. The researcher's inability to adequately distance themselves from respondents during the data collection process might jeopardise the reliability of the study. This is essentially a philosophical issue that calls for a philosophical perspective when discussing indigenous subjects, particularly in postgraduate research.

Philosophical thinking about research and knowledge production

Concerning novel methods of thinking, Francis Bacon suggests that scholars should avoid making assumptions based solely on speculation. This is due to the fact that deductive thinking, which is prevalent in postgraduate study, can be troublesome and speculative (Bacon, 2022). It implies that postgraduate research students, especially those in STEM, may not be permitted to think beyond content knowledge, which could impact the quality of their outputs even in their future professions (Ding, Wei, & Mollohan, 2016). Postgraduate students tend to reach several conclusions based on what they have read or experienced before the start of their studies. Precociously reaching a conclusion can affect their judgement in dealing with data related to indigenous issues, for instance. This inability to deduce problems may lead to answering research questions for participants or even influence how the participants answer those questions. In other words, the researchers already have an idea of the problem and what needs to be reported. If the research problems are framed in this manner, the indigenous methods of inquiry are likely to produce emotion-laden data and not information based on the participants' reality. This type of thinking may also affect how the problems are framed and shown in the literature. While Bacon's idea is that human experience can be the basis for the scientific validation of knowledge (Remmling, 2022), this can also be problematic in formulating research problems for postgraduate students as it may lead to formulating problems from personal experiences and practices which may not be valid or out of context. Therefore, Bacon and pragmatists in postgraduate research and supervision in higher education strongly suggest not producing knowledge solely on the human experience. As such, Bacon's recommendation to use the experimental study to engage ideas and arrive at an effective system of knowledge (Corneanu, 2011) is valid. Still, it can also be challenging because not all problems in higher education research require experimentation or can be observed, such as issues about indigenous people in Africa and elsewhere. The idea that knowledge generated outside this process is falsehood is problematic and not pragmatic.

Therefore, an inductive approach in pragmatism (Liu, 2016) will allow us to discover research problems by reading and mainly through socialisation in the learning culture. This may require providing a conducive environment that enables supervisors and students to develop collegial relations where it neither imposes any form of thinking but allows a knowledge generation process from general observation of phenomena before concluding on topical issues. Therefore, the inductive approach could also mean bringing experience into research because personal meaning could provide a useful foundation for the truth about researching topics that are complicated or complex in higher education research. In this way, the aim is to debate whether knowledge can come from nothing more than one's own personal or intellectual experience. Nonetheless, the core principles of pragmatists' educational philosophy are based on personal experience and meaning (Sharma, Devi, & Kumari, 2018). If this approach is used to explore indigenous subjects, mistakes are likely to be made in the data collection and interpretation process.

John Locke

In this context, John Locke (1632–1704) tried to prove that humans can inadvertently conceive wrong ideas from childhood (Leibniz, 2017). Regarding research related to social issues, the notion of conception from birth can be transformed by what a child sees in the environment. This also can influence perception about other feelings. Such ideas were spread through stories told to children about how others perceive them in society. Youngsters frequently struggle to understand complicated matters and may conclude about the world based on what adults have told them. For instance, a story about the injustices during colonialisation may contribute to children developing resentment towards those who colonised their ancestors. In many cases, it has led to violent behaviour, which is why knowledge about colonial issues is still challenging to unravel. Reflections and comparisons of ideas are essential for knowledge validation. However, human beings often refuse to critically analyse knowledge adopted in childhood owing to cultural or religious beliefs. In terms of research, students with pre-conceived ideas may struggle to unlearn preconceptions and acquire new knowledge.

Postgraduate research and studies, if the concept is to be followed, should help students understand concepts prior to graduation. The aim is not to dictate the concepts to be used to consider particular issues. An approach enabling the study of many epistemologies is required for historical issues relating to race, colonisation, and other complicated topics. How can students and supervisors ensure that what they have taught or experienced will not influence the design of their studies? African epistemologies are likely to be challenges for propagating ideologies that cannot be proven because their foundation rests on feelings and stories. In other words, African knowledge may need to discover new approaches to prevent biased research and findings. While students should be independent in thinking, this can also be guided to ensure that African views on sensitive issues are holistic and argued on theories that are not controlled by human minds.

Jean-Jacques Rousseau

Human minds can be swayed by unverified ideas that seem genuine. These ideas are developed through parental training and religious exposure. The flexibility of new knowledge becomes challenging for an individual whose mind is closed to new ways of producing knowledge. At times, students would like to change supervisors for non-academic reasons: simply because they do not share the same worldviews. However, Jean-Jacques Rousseau (1712–1778) condemned civilisation and blamed it for distancing humans from nature, which was supposed to help generate new ideas or knowledge (Geoffroy-Schwinden, 2014). Rousseau believes that being natural is an essential aspect of pragmatism (Giladi, 2019; Kaushik & Walsh, 2019), but civilisation caused by technological innovations should be contained to allow human minds to function correctly. Therefore, naturalism in higher education, research, or postgraduate supervision means not controlling students' minds but shaping their thinking. It is unnatural to restrict postgraduate students to one way of researching issues in higher education.

Thus, knowledge is nature-based and closely linked to the human experience. For instance, in choosing research topics, students should be allowed to conduct a study connected to their meaning and experience. Some students have been subjected to abuse and all forms of discrimination before joining the university. Such students may be looking for answers through their research design. This could influence the interpretation of existing literature. Therefore, such students may select a topic related to stories they have been told or related to their own experiences. Students may sometimes like to research the link between the history of Africa and the role of women in African society. This topic often reflects what the students have seen in their households or personally experienced. However, connecting such research to the appropriate philosophical foundation to generate sound knowledge may be challenging unless objectivity is observed. Thus, postgraduate research goes beyond studying a social phenomenon, for example. Many human elements could influence the knowledge production process. The nature and natural experiences of researchers play a significant role in creating knowledge that is not necessarily based on science.

August Comte and Charles Sanders Peirce

Furthermore, August Comte (1793–1857) puts forth that science solves problems through empirical inquiries or experiments which have generated scepticism (Heilbron, 2017; Hoecker-Drysdale, 2014). On the one hand, the belief in science as a panacea presents an issue. On the other hand, modern civilisation encourages a scientific approach to pragmatism and solving societal problems. Whether science provides ultimate ways of knowing, existence, and structure needs further investigation in higher education research. Charles Darwin (1809–1882) is opened in his view of knowledge creation and pragmatism because it challenges the belief that the development of ideas is the concern of a supernatural being (Prázný, 2021). This view in researching religious topics is problematic for students and supervisors with different experiences. Likewise, Charles Sanders Peirce (1839–1914) rejects the conception of ideas in a pattern and refuses the dualism of the mind about reality or truth (Hoopes, 2014; Ibri, 2017). Therefore, how would a study test experiences of past racism, discrimination, and oppression by simply asking for stories from people who have experienced them? There might be a need to revalidate stories and find alternative methods to ensure that the data collection process is not merely reporting what is intended in the minds.

The idea of testing and validating knowledge is popular in postgraduate research. This knowledge creation might not improve our understanding of topics where experimentation is unnecessary but where human accounts and other evidence must establish facts. Both the subject of race and religion are contentious. As such, it may be necessary to contest the notion that science is needed to reach and prove the truth in an African context. In this case, it might be interesting to generate other ways or methodologies to analyse this problem without necessarily testing and conducting an experimental study. The truth about African issues is embedded in its history and is not necessary in testing.

William James and John Dewey

On the other hand, William James (1842–1910) believes the truth to be made through real-life events (Radu, 2019). With this, students may design the study so that the data are collected by observing and analysing truths about the investigated problems. This may include going to the sites where an incident occurs in the past, examining colonial objects, or revalidating the truth about topics of oppression and the history of colonialism. This also depends on whether the researchers can objectively provide universally acceptable knowledge since truth does not belong to ideas (Shook, 2011). Therefore, the onus falls on researchers to accurately demonstrate how objects are analysed beyond researchers' views. How would researchers ensure that the truth is discovered based on facts validated through different sources? Likewise, truth is not always objective, verifiable, or verified objectively (Zhao, 2014). In that case, who's true?

Thus, African epistemologies face different perspectives on how truth can be generated and validated. African science will continue to be undermined if they are based on the scientific way of arriving at truth. The point is that Africans' true history and experience can be used to justify the truth about problematic issues facing African societies. Experience is of nature (Rockefeller, 2014). Therefore, truth is inseparable from experience (Lapoujade, 2019) because the universe is open-ended, pluralistic, and in the process (Leary, 2010). Therefore, John Dewey (1859–1952) made pragmatism popular and offered the idea that experience reaches into nature and is not two different things (Alexander, 2012). Yet, the African experience is still not recognised as valid epistemology.

The following section offers different ways of thinking about knowledge and research in higher education from different knowledge domains. The emphasis is placed on social science research. In this case, students will confront ideas and thinking that will challenge the literature and the research team. Unlearning colonial ways of conducting research in Africa requires providing alternative evidence that is universally acceptable. This section argues whether the African experience of the past is an accurate methodology that can be justified as a valid way of knowing.

African and non-Western worldview in research

It has been suggested that social scientists should leave philosophical musings to African philosophers and concentrate on translating the ideas into actions to guide people with respect for African or non-Western worldviews in their research undertakings. This idea is valid for social science researchers because Africans have unique experiences that have formed their views on specific topics and issues. This is different for non-African social scientists because they do not have any lived experience of the problems being researched. Therefore, asking African students to think globally may be confusing for some social issue topics that may deny students the opportunity to think about their problems and how they would like to interpret them. This can similarly be blurred by Western theories and methodologies prescribed by supervisors who may struggle to understand students' perspectives. The only way to prove African knowledge is to allow flexibility of independent thinking.

In most cases, the universities create academic-student tensions because research projects sometimes already designed before students are recruited into them. This is becoming a battlefield for students to constantly show that the existing framework of creating knowledge needs alternative epistemological options. The conflict is between the mind and the inability to unlearn colonial ways of studying African problems that are uniquely different and beyond science. One way to look at this tension is to challenge colonial epistemologies by offering new ways of thinking and doing research that is concerned with African histories. The idea that science solves all problems is also problematic. Meanwhile, there is another dimension to this problem. Other African researchers will like to keep Western epistemologies or refuse to adopt African thinking to study social issues pertaining to African people. This creates epistemic violence and injustice among Africans who want to solve African problems using only Science. A significant issue is African knowledge being deemed 'not modern' and requires the premise of science for validation purposes. As it has been established, science researchers question to obtain measuring observable data on variables. The main statement of purpose is scientific, and the hypotheses – and the research questions are precise and close so that the researcher can separate a few variables for the study. In studying these variables, a researcher can get measurable or assess data on an instrument or a scale. The challenge is whether this process can be applied to solve all major issues affecting Africa, like racism, violence, terrorism, and ethnic and multicultural problems in classrooms, without considering Western theories or epistemological pathways to arrive at the truth about all these issues. The argument is that not all problems or issues can be separated into variables, counted, or have a causal relationship. There are problems whose solutions lie in telling stories about who understands its history with certain evidence that may not necessarily follow the positivist philosophy of knowledge.

Africa is a large continent with a large population. It has archaeological evidence that knowledge about the world originated from the continent, including the first religion and method of governing humans. How did Africa realise other forms of knowledge and abandon its history and cultural beliefs? History shows that methods of thinking about issues were imposed on early Africans. Force, in this way, means subjecting their minds and emotions to fear. Africans have been cultured to see their beliefs and knowledge as evil or non-civilised. For example, African knowledge of supernatural powers is termed 'evil' but can be used to reveal forgotten stories of African civilisations. Fast-forward, the issue remains whether African knowledge is acceptable in the modern world. Interestingly, several Western countries are now learning African histories and languages. One would expect that this development will bring about an appreciation of African epistemologies, not just learning about past events. However, it seems a struggle exists to put forward African ways of knowing in an acceptable global domain of knowledge. In other words, the modern era still wants to control global knowledge through the narrative of different forms of knowledge that are not global.

Who controls the creation and spread of knowledge in the twentieth century is evident in the growth of migration. There is a concern that the only practical and universal knowledge comes from outside of Africa. Universities in Africa must compete using the Northern knowledge dimension as their framework. Although African knowledge may not be pertinent to their situation, a few American and European students seem eager to travel to Africa to study. It shows the power and assertiveness of the West to put Africa and its scholars behind in terms of their ability to demonstrate knowledge of various matters. When Western students or scholars visit Africa, the visit is based on the narrative to help Africa grow their research and overall knowledge by sharing and promoting Western approaches and knowledge. This indicates why study exchange concerning globalisation and internationalisation equates to Africans learning from the West to return to Africa to propagate the acquired Western knowledge. Knowledge is thus understood in light of the North's dominance and its potential future impact on the minds of Africans. Due to their inadequate English language proficiency or use of an African accent when using the colonial language of English, African students may opt to learn to speak like Americans or feel proud to employ a British accent and may even discriminate against other Africans. The colonial era, when the so-called universal language could only convey what we already know, was revived by this movement. However, African heritage, culture, and language cannot see its value.

Concluding thoughts

Research and knowledge in Africa are for social and national purposes. The knowledge produced in Africa has always been in the public domain and subject to public scrutiny. Meanwhile, knowledge is a central societal agency with social and political consequences. African researchers, including students, are subjected to social, economic, and political pressures to prove what they know. Researchers now find themselves in situations that are no longer solely about knowledge production. Still, they must show that African ways of thinking are valid to study complex issues. Contemporary researchers not only have to deal with questions of rejecting scientific methodologies but also be concerned with issues related to showing African histories as one of the principal methodologies to understanding complex problems in Africa and globally. Higher education is a social institution and should be an agent of social change where knowledge is not controlled or prescribed. Those adherent to this concept were the organisers of the progressive knowledge movement in Africa. They usually take the dynamically changing universe as their frame of reference. Since nothing can be accepted as final in a world ruled by flux, they taught their students little about what to think in the research process. Students learn to regards conclusions as tentative and subject to amendment in light of future events. By introducing this into postgraduate research, progressive knowledge producers kept a sharp watch for signs indicating the probable future direction of social change. It is believed that higher education should be a place where all knowledge and methodologies are allowed, no matter how varied or contradictory they might be, and institutionalise the continuous reconstruction of social theory and practice. Every knowledge enjoys considerable autonomy in determining its own character. Therefore, higher education, although maintained by the state, should not be its supine servant. On the contrary, it should be free to determine its own function in the social order.

References

Ahluwalia, P. (2012). *Politics and post-colonial theory: African Inflections*. New York: Routledge.

Alemu, S. K. (2018). The meaning, idea and history of university/higher education in Africa: A brief literature review. *FIRE: Forum for International Research in Education, 4*(3), 210–227).

Alexander, T. M. (2012). *John Dewey's theory of art, experience, and nature: The horizons of feeling*. New York: SUNY Press.

Bacon, F. (2022). Francis Bacon's natural philosophy and medicine. *JMS*. https://hekint.org/2022/09/29/francis-bacons-natural-philosophy-and-medicine/

Björkman, B. (2018). "This is not familiar to most people": Navigating peer reviewers' comments and knowledge construction practices by PhD students in supervision interactions. *Journal of English as a Lingua Franca*, *7*(2), 333–354.

Corneanu, S. (2011). *Regimens of the mind: Boyle, Locke, and the early modern cultura animi tradition*. Chicago, IL: University of Chicago Press.

Davis-Warra, J. (2017). *Durithunga—growing, nurturing, challenging and supporting urban Indigenous leadership in education* (Doctoral dissertation, doctoral thesis, Faculty of Education, The University of Queensland).

Dewey, J., & Imagination, M. (2020). Pragmatism and ethics. In *Works Righteousness: Material Practice in Ethical Theory*, 65.

Ding, L., Wei, X., & Mollohan, K. (2016). Does higher education improve student scientific reasoning skills? *International Journal of Science and Mathematics Education*, *14*(4), 619–634.

Geoffroy-Schwinden, R. D. (2014). Rousseau and the revolutionary repertoire. *Studies in Eighteenth-Century Culture*, *43*(1), 89–110.

Giladi, P. (Ed.). (2019). *Responses to naturalism: Critical perspectives from idealism and pragmatism*. New York: Routledge.

Gullbekk, E., & Byström, K. (2018). Becoming a scholar by publication–PhD students citing in interdisciplinary argumentation. *Journal of Documentation*, *75*(1), 247–269.

Heilbron, J. (2017). Auguste Comte and the Second Scientific Revolution. *The Anthem Companion to Comte*, 23–42.

Hoecker-Drysdale, S. (2014). Harriet Martineau and the positivism of Auguste Comte. In *Harriet Martineau* (pp. 187–208). New York: Routledge.

Hoopes, J. (Ed.). (2014). *Peirce on signs: Writings on semiotic by Charles Sanders Peirce*. Chapel Hill, NC, and London, England: UNC Press Books.

Ibri, I. A. (2017). *Kósmos Noetós: The metaphysical architecture of Charles S. Peirce* (Vol. 131). Cham, Switzerland: Springer.

Kaushik, V., & Walsh, C. A. (2019). Pragmatism as a research paradigm and its implications for social work research. *Social Sciences*, *8*(9), 255.

Kelly, L. M., & Cordeiro, M. (2020). Three principles of pragmatism for research on organizational processes. *Methodological Innovations*, *13*(2), 2059799120937242.

Lapoujade, D. (2019). *William James: Empiricism and pragmatism*. Duke University Press. Durham, North Carolina, United States.

Leary, D. E. (2010). Instead of Erklären and Verstehen: William James on human understanding. In *Historical perspectives on Erklären and Verstehen* (pp. 121–140). Dordrecht, the Netherlands: Springer.

Leibniz, G. https://www.earlymoderntexts.com/assets/pdfs/leibniz1705book1.pdf

Liu, L. (2016). Using generic inductive approach in qualitative educational research: A case study analysis. *Journal of Education and Learning*, *5*(2), 129–135.

Oppong, S. (2013). Indigenizing knowledge for development: Epistemological and pedagogical approaches. *Africanus*, *43*(2), 34–50.

Prasad, M. (2021). Pragmatism as problem solving. *Socius*, *7*, 2378023121993991.

Prázný, A. (2021). New education and Deweys Pragmatism. *Editorial & Advisory Boards*, 12(1), 91.

Radu, M. (2019). William James: On the border between sciences and religion. *Euromentor Journal-Studies About Education*, *10*(2), 26–38.

Remmling, G. W. (2022). Francis Bacon and the French enlightenment philosophers. In *Towards the sociology of knowledge* (pp. 47–59). London, England: Routledge.

Rensfeldt, A. B. (2013). Imaginaries of "Europe" in the governmentality of PhD education. *European Education, 45*(4), 44–60.

Rockefeller, S. C. (2014). John Dewey. In *Twentieth-century philosophy of religion* (pp. 39–52). London, England: Routledge.

Sharma, S., Devi, R., & Kumari, J. (2018). Pragmatism in education. *International Journal of Engineering Technology Science and Research, 5*(1), 1549–1554.

Shook, J. R. (Ed.). (2011). *The essential William James*. New York: Prometheus Books.

Stilwell, C. (2010). Understanding indigenous knowledge: Bridging the knowledge gap through a knowledge creation model for agricultural development. *South African Journal of Information Management, 12*(1), 1–8.

Vavrus, F., & Bartlett, L. (2012). Comparative pedagogies and epistemological diversity: Social and materials contexts of teaching in Tanzania. *Comparative Education Review, 56*(4), 634–658.

Whyte, M., & Whyte, S. R. (2015). Dilemmas of knowledge production in Ugandan universities. In *Higher education and capacity building in Africa* (pp. 41–57). London, England: Routledge.

Zhao, S. (2014). Self as an emic object: A re-reading of William James on self. *Theory & Psychology, 24*(2), 199–216.

4 Epistemic injustice in research publication

Introduction

The relevance of knowledge and knowledge production control are post-colonial justice and neoliberalist issues (Contu, 2020). Despite applying post-modern thoughts in global knowledge studies, postmodernists ignore the realities of African scientists or scholars from the Global South (Tripathi, 2021). The critical question is whether the editorial review process for African publications mimics colonialism's structure. This issue requires deconstructing the academic imperialism (Patel, 2022) concept from an anthropological lens to argue knowledge inequality in research publications in Africa cultural. *Cultural* in the sense is the culture of power and belongingness that still drives academics' orientation about knowledge production due to their desire to publish in so-called international journals. A historical approach is necessary to critique the postmodern transformation of knowledge's purpose in an African cross-cultural context (Higgs, 2017). While it is essential to be critical of neo-liberal and post-colonial politics in knowledge production (Parashar & Schulz, 2021), power relations in knowledge production can be analysed to explain Africa's low impact on global research outputs.

Whether the editorial review process for African publications mimics colonialism's structure is critical to studying African epistemologies. Academic imperialism and anthropological analysis are critiquing forces of globalisation, knowledge globalisation, and social betterment through epistemic justice (Rudolph, Sriprakash, & Gerrard, 2018). Postmodern transformation is either critical or sceptical of modernist views or epistemic certainty since the premise is that there is no fixed truth about knowledge creation and knowledge standardisation (Gojkov, 2012; Watson, 2020). The idea of truth's fluidity is related mainly to Foucault's concept of knowledge and power in relation to the purpose of knowledge (Ahmeti, 2021). Regarding postmodern thoughts in global knowledge studies, something can be learnt from Thomas Aquinas's critique of Aristotle's work (Günther, 2016). The theological stance of Catholics in Rome challenged Aristotle's idea about knowledge. This led to changes in the general belief about religion and knowledge about God. Thus, scepticism is the first stage of challenging and creating other forms of knowledge, including in higher education research.

DOI: 10.4324/9781003364740-5

Knowledge production in higher education seems to follow neo-liberal and post-colonial politics. This market-oriented approach is the reason for the crisis in higher education. Knowledge is seen as a countable entity or a commodity since outputs are tied to specific rewards, such as promotions. This philosophy about knowledge production creates unhealthy competition among academics and leads to the proliferation of primarily low-quality research outputs. Arguing this issue does not consider quantitative outcomes or the number of research outputs because counting knowledge is one of the features or problems of managerialism and neo-liberal policies in the governance of higher education systems (Zajda, 2020).

Whether African scholars have faced prejudice during the editorial review of their articles can also be examined from the viewpoint of quality. This is a complicated question; however, it is presumed that academics wanting to publish in high-impact academic journals should take advantage of the objective peer review process to raise the quality of their work before publication, regardless of the manuscript's origins. Therefore, rejection may not necessarily equate to discrimination but is simply an opportunity to enhance researchers' thinking and writing skills. Supporting this claim can be challenging as some researchers believe their knowledge is not recognised and published because of their skin colour, not necessarily for quality.

On the other hand, focusing on counting knowledge and research production speed in academic paper submissions is deeply problematic (de Guevara & Kostić, 2017). Arguments about knowledge production in the context of journals and racism draw attention to power and politics around knowledge dissemination in higher education. Therefore, the literature on colonialism, the history of Africa, and knowledge can be reviewed to analyse ideas on the trends in knowledge mobility and the dimensions of knowledge-sharing processes. Meanwhile, southern theories on this issue can bring an important perspective to understanding the dynamics of knowledge transmission and its complexity in the African context. The chapter argues that Global South epistemologies and policies can be better understood by investigating present sociologies and histories shaping knowledge sharing in Africa. Thus, it seeks to answer questions about how knowledge creation and sharing can better align with African development. Central to the chapter is determining what the globalisation of knowledge means in the context of Africa and its relation to epistemic injustice.

Globalisation of knowledge

Joseph Nye and John Donahue define globalism as a world state involving networks of interdependence at multi-continental distances. In higher education, globalism means collaboration, partnership, and sharing knowledge regardless of the location of the institutions. For Africa, globalism pertains to deciding the area of cooperation and institutions for collaboration regarding knowledge production. In that case, African higher education institutions are proud of cross-border interaction with top universities in the West, as they may not want to partner or establish strong ties with institutions at the regional level. Beneficial association with the West and distanciation from African academia are African scholars' understanding of globalisation and knowledge globalisation. Simply put, establishing solid ties with Western higher education institutions for recognition, not the other way around, is the objective of knowledge globalisation.

International development organisations promoting knowledge production on issues of poverty in Africa need local relevant knowledge on the problems. Discussing such topics from a Global North perspective and without many regional collaborations can be contentious. Findings often lump a whole continent together as one country and only project Africa as a place of poverty. Therefore, different dimensions and knowledge about poverty are wrongly shared in reports often produced by international development organisations such as the World Bank, IMF, and similar structures. Though these organisations collaborate with African institutions, data in the reports suggest a lack of knowledge in discussing African matters. The migration of knowledge should improve the world's economy through properly shared knowledge.

Likewise, Thomas Friedman's idea of globalisation is of the Lexus and the Olive Tree. This idea speaks to a globalist system that has replaced the Cold War system. The new system has unique rules, logic, pressures, and incentives driven by international capitalism. The feature of integration is the free flow of capital, goods, and ideas more broadly, faster, and more profoundly than ever. In other words, abiding by free market principles and efficiency goals is essentially a 'Golden Straitjacket'.

For higher education and knowledge production, Africa must determine whether speed and capitalism make sense for its system. In this way, assessing the kind of knowledge to exchange with the world and determining whether conforming to international capitalist systems is plausible for Africa is critical. African discoveries, such as scientific solutions to global problems such as the South African discovery of a new COVID-19 strain in South Africa, were seen as a problem created by African scientists rather than a significant concern. Notably, though the COVID-19 pandemic did not originate from the African continent, its solutions cannot only be from the Global North as it is a global predicament. Yet, African countries were banned from discovering the omicron strain instead of recognising the knowledge as an effort towards solving the global pandemic. First, this displays knowledge imperialism aiming to place Africa at the bottom of the global knowledge production echelons and extend Global North power, control and dominion. Second, capitalist ideologies aim to ensure the disregard of African discoveries, including vaccines. Finally, through globalisation and capitalist ideals, the West seeks to grow its economy and further establish itself as a superpower in all spheres, including knowledge production. Thus, globalisation and knowledge globalisation in Africa must be rethought.

Players in globalisation objectives also promote cultural homogenisation in knowledge acquisition and sharing. According to Friedman, the United States is paramount in globalisation's new structure and balance of power and the traditional balance between states (countries). For example, achieving a balance between the United States and global markets regarding knowledge sharing involves the recognition of the African market by the Global North. Ground-breaking knowledge emanates from the African continent with the potential to change the world.

A balance between individual academics and institutions is the knowledge that influences governments' decision about accepted or rejected.

In many cases, the Global North often influences African politics and governments through funding, determining the focus of the knowledge expected from Africa. Hence, globalisation produces 'super-empowered individuals in the Global North that can control knowledge in other countries, including outlets where the knowledge is disseminated'. Consequently, many journals in the Global North are empowered by their boards and editors to ensure they serve as gatekeepers for knowledge that is against their agenda. The crucial conclusions from Friedman's argument are reward, peace and democratic dividends. First, reward refers to the incentives given to those participating in globalisation. In other words, winners and losers are selected for this type of participation. Second, a peace dividend means that those engaged in globalisation have too much to lose should things go wrong, as with epistemic violence. Africa, in this case, has lost value and the ability to operate confidently in the global dissemination of knowledge. Finally, democratic dividend refers to the free movement of information. In reality, this is contained and controlled by journals and represents another means of reinforcing the power of controllers of knowledge or global economics.

That is not to say that globalisation is evil. Still, it should be noted that the advantages of globalisation depend on a collegiality that can also recognise knowledge coming from Africa for development purposes. For instance, economies in post-industrial societies increasingly rely on the service sector. Though Africa is included in globalisation endeavours, exploitation rather than global benefits was the primary motivator of this inclusion. In the same way, higher education tends to follow a market approach to grow capitalists' power further. As a result, this has promoted managerialism and capitalism ideologies in higher education systems. While some countries are challenging the status quo due to its adverse effects on quality research, other countries, especially in Africa, have welcomed the establishment of power relations that allow corruption in higher education and the opportunity to dictate rather than lead scientific research and knowledge.

The following section concerns neoliberalism and its influence on research publications in the higher education system. Knowledge production in higher education is encouraged to move away from a business orientation to an intellectual enterprise. A shift in focus will allow the prioritisation of research and knowledge production for the empowerment and transformation of society rather than profit-making or wealth accumulation. Though higher education relies on capital generation, including knowledge produced by academics for sustainability, this idea has caused academic tension regarding workload and performance management.

Implication of neoliberalism in knowledge production

In terms of knowledge production, Lynch's (2015) writing speaks to new managerialism, neoliberalism, and capitalism as social theories in higher education. In higher education systems, the relationship between parties involved in knowledge production and management is social. However, such a relationship has to be both social and intellectual when producing knowledge. In other words, the university's duties are to create an environment suitable for knowledge production since academics are expected to use the facilities to develop knowledge.

On the other hand, the interaction between academics implies a mutual understanding between those responsible for producing knowledge. For instance, increasing academics' workload to meet targets may affect mental health and social relations in higher education. In most cases, burnout experienced by academics leads to absenteeism and conflicts with managers who treat higher education as a factory. A positive relationship is unlikely to be achieved at this stage and could lead to high resignation rates. Meanwhile, academics who care less about quality and can play the political game may stay in a complex social environment where their health is sacrificed for the speed of research publication.

The marketisation of higher education started in the 1980s and impacted education (Lynch, 2014). Meanwhile, academic capitalism began in the United States in the 1940s and has been studied for the future of Australian higher education (Croucher & Lacy, 2020). Interestingly, this story and history show that the marketisation of higher education emanated from the Global North and migrated to different parts of the world. It is now also reflected profoundly in the exchange of knowledge globally. The moral aspect of the new managerialism and its carelessness in higher education is a concern (Lynch, 2010). In other words, intellectuals are not the decision-makers of higher education management because they believe themselves incapable of generating income for the sustainability of the systems. Hence, managers who give instructions and sometimes threaten to withdraw rewards in case of insubordination or non-compliance have taken charge of the university's affairs. Vican, Friedman, and Andreasen (2020) speak on responding to the pressure posed by neoliberalism concerning major issues in higher education. This pressure includes meeting targets and fast knowledge production to benefit from government subsidisation, which has influenced and shaped the future of higher education (del Cerro Santamaria, 2020).

The marketisation of higher education is now becoming popular in Scandinavian countries (Alajoutsijärvi, Alon, & Pinheiro, 2021). Whether Europe thinks knowledge creation in higher education should be equivalent to private goods for economic development is one thing. Another thing is how to produce quality using the industry-based methodology of neoliberalism in higher education settings. This is a long-withstanding concern in Africa. Development in Africa and elsewhere is linked to the quality of the knowledge produced. Western philosophies of knowledge production and dissemination are often duplicated in Africa. Therefore, the West's endorsement and adoption of neoliberal ideas may influence Africa's choice to adopt the aforementioned due to the belief in Western epistemologies' ability to validate African knowledge. The application of neoliberalism in Western higher education may be different in terms of their context and different knowledge domains. If this is not considered in Africa, the market may slowly engulf the future of African higher education and knowledge. This is the reason Bessant, Robinson, and Ormerod (2015) call for a rethink and to consider the higher education space as a radical democratic social place. Discussions around these issues are critical among different parties and stakeholders to create new knowledge creation systems without losing quality. Likewise, such debates should assertively challenge corporate influence over higher education and neoliberalism's trigger of knowledge marketisation (del Cerro Santamaría, 2021).

Similarly, Peters (2020) uses the knowledge capitalism construct to argue that research publication processes are becoming compliance exercises and not necessarily quality knowledge production processes for societal development. Hence, neoliberal and managerialist ideologies and Keynesian Theory in relation to the purpose of higher education are deeply problematic (Nureev, Volchik, & Strielkowski, 2020). Thus, the issue of quality, quality assurance, and neoliberal ideas in higher education need reforms (Raaper, 2016).

In higher education management and knowledge production, Broucker, De Wit, and Verhoeven (2017) articulate the neoliberal approach and academic capitalism as the new public manager in higher education. Likewise, del Cerro Santamaria (2019) critiques the prevailing model of capitalism in defining the meaning of higher education and its implications for the labour market economy. Besley and Peters (2009) explain how neoliberal and managerialist policies have been restructuring and reforming their higher education systems. Lynch (2017) speak about performance and capitalism in defining higher education's meaning. Lynch (2012) discusses global capitalism's control of education policy globally. Tight (2019) calls for a rethink of the concept of students as consumers because it propagates managerialism in higher education systems. To challenge this notion, Hanesworth (2017) argues that academic capitalism and policy are necessary to respond to managerialism's effect. To ensure that changes are not superficial, McClure (2016) suggest knowledge privatisation influences the objective of universities.

Several countries are already changing their views about capitalist and neoliberalist ideologies in higher education systems. For instance, Mirzamohammadi and Mohammadi (2017) provide information regarding capitalism and neoliberalism in Iranian higher education and how the country has opposed these ideologies since the Islamic revolution in 1979. Erickson, Hanna, and Walker (2021) critique how neoliberal universities allow specific knowledge. This may be the reason to challenge it more. In other words, managers are likely to prescribe the type of knowledge and knowledge dissemination required by the economy. In this case, academics are not allowed to make decisions about intellectual matters or contribute to knowledge because of different forces serving as barriers. So, Narkunas (2020) believes that universities in favour of capitalism and neoliberalism ideologies do so to benefit from a return on their investment. In other words, every academic is an investment unit; if the knowledge they produce cannot financially sustain the university, it is not 'good' knowledge. As such, Ivanova Kremakova (2016) argues that neoliberalism turns higher education into factory workers and calls for radical change to minimise its effects. This is how new public management and new managerialism work (Dougherty & Natow, 2020).

Furthermore, it explains how they are shaping the operation and governance of higher education. Jackson (2017) calls it neoliberal globalisation, which leads to new forms of imperialism and division of power for historically disadvantaged groups in societies, including causing epistemological barriers. Rowlands (2017) argues against the globalisation of knowledge and policy. Wood (2010) calls the entire process of changing knowledge space in higher education a neoliberal fantasy and a problem of making knowledge creation a business enterprise. Finally, McCann, Granter, Hyde, and Aroles (2020) think the marketisation of higher education makes leadership more autocratic, which takes away freedom of speech, academic freedom, and democratic participation on campus. The approach has increased the workload in British universities and applied pressure on academic staff to perform (Feldman & Sandoval, 2018), similar to events in African higher education. This means that global efforts are needed to rethink how universities can break from ideologies affecting knowledge production and damaging the social connections of academics responsible for intellectual activities. Higher education is supposed to be a social good where different stakeholders interpret and implement policies for development purposes, but this is debatable (Connell, 2017). Regardless, what counts as knowledge can be explored and examined outside academic capitalism, promoting the marketisation of higher education (Sigahi & Saltorato, 2020). In conclusion, modern and neoliberal universities have made determining knowledge more difficult (Klikauer & Link, 2021).

The following section speaks to the attribution of knowledge and who has the power to determine its production. This is discussed in terms of claims that African researchers have been discriminated against for their knowledge. Could this relate to how universities are managed or other issues affecting the modern university? This section explores different matters around the editorial review process in publications.

Power and politics in research publication

African theory of knowledge, marketisation, and commercialisation of knowledge shape post-colonial thinking and research in higher education. Appraising African identity in global knowledge production is vital to evaluating the knowledge relevant to the African context. Knowledge marketisation and commercialisation have become popular in Africa and around the world. This is simply due to competition for funding and recognition. The dimension of this problem in Africa is the reason for poor quality research publications and several publications of low-impact factors.

Another issue is the criteria used in the review of the knowledge produced by Africa and its treatment in foreign journals need to be assessed. This question challenges injustice by checking whether journals review African knowledge, universal knowledge, or Western epistemology, which needs to be proven by the journal's editorial policy. How do we establish that the editorial review process is biased against African researchers? Who can authenticate the reliability of the editorial review process of papers from Africa? Reasons behind the rejection of African research papers include inadequate research training, language, method of research, and others. One wonders who has the power and knowledge to write about topical issues in higher education.

The following section argues for African ways of knowing and knowledge production in the African context. The examples and situations of likely injustice for African researchers based on their language and knowledge and in submitting academic papers for publication are discussed. The discussion claims that there might be a form of justice, but they are not limited to the peer-review process of scholarly journals. This also includes policy problems around issues of publication in Africa. The conclusion is made to rethink a policy seemingly discriminating against indigenous knowledge and African ways of knowing through policies from African higher education systems based on the philosophy that knowledge must be counted and produced fast. There is a need for reflection on the colonial discussion of African epistemologies to show possible options for improving research output and performance without necessarily ignoring essential collaboration and partnership with scholars and organisations in the Global North.

Colonial discourses of African epistemologies

Colonial discourses of African ways of knowing are topics of interest for some scholars (Ngara, 2007). Black African science, indigenous knowledge, and Western epistemology have been discussed in the literature (Luyaluka, 2016). The literature has also featured African epistemology and the idea that science is rational (truth based on evidence), universal, and methodological (Alemu, 2019). Discrimination can occur in the peer-review process with papers originating from African countries. A recent publication in World Universities News by Wagdy Sawahel on 27 January 2022 indicates no evidence of bias in the peer review process against African research. However, according to the opinion of a few editors of African journals, the report reveals unconscious bias. The argument is that the focus should be on strengthening African research capacity.

Determining bias is difficult, but the peer-review process can be strengthened to counter unconscious bias in the review process. Likewise, publications by scholars such as Medie and Kang (2018) called attention to several problems related to what most colonial researchers have called academic imperialism and/or injustice in knowledge production. For instance, a large majority, or 85% (Medie & Kang 2018) of the editorial board members of top journals are from the United States and Europe and determine the truth and validity of knowledge dissemination in science. Less than 3% of articles in four leading journals were by writers in the Global South (Medie & Kang 2018). In other words, African researchers represent approximately 3% (Medie & Kang 2018) of the total percentage of academics sharing their knowledge in these journals.

Interestingly, South African and Indian universities constitute these 3%, leaving out researchers and universities from other African countries. This speaks to capacity in terms of the ability to deliver quality, not necessarily because of gatekeeping reports that usually happen among journals not situated in Africa. In this argument, we try to show that capacity and resources are not the reasons for academics' production of quality research. Yet, capacity and resources are vital for academics to produce high-quality research. Though this is an assumption, it is not likely that quality research will be rejected. In another sense, most well-resourced universities are reputable and have an international collaboration that allows them to publish in high-impact academic journals.

Likewise, findings show that funding, government policy, and poor infrastructure have affected African researchers and research production (Medie & Kang, 2018). The few Africans able to publish in high-impact journals have experienced other forms of challenges and injustices (Da Silva & Kimotho, 2022), such as language, nationality of the researchers, and rejection based on the impression that African publications may be inferior. This is not easy to prove, but it is mostly the feeling of many African researchers based on the feedback they often receive from editors of journals in North America and Western Europe. Moreover, policies on publication in some African universities are problematic and have either discouraged academics or did not fund publications in local journals, even if the manuscript's content is irrelevant to foreign journals. This is another form of injustice and discrimination that now comes from within and among Africans themselves. So, what does this means for the African Theory of knowledge and critique of the history of colonialism, including colonial discourses (Ngara, 2007)? The main argument is that Africa's low impact on global research is due to two forms of injustice. One is the injustice that could be due to the editorial review process of most journals in the Global North. The second is the top-down policy approach that still mimics colonialism's policies in Africa.

Reflecting on Robert Jensen's idea on the Theory of knowledge, he explains three theories of truth: correspondence, coherence, and consensus. Correspondence theory is about justifying knowing or truth. This thinking speaks to African knowledge claims of the experience of discrimination in publication, but such claims cannot be regarded as true unless they can be justified. Otherwise, the claim will be termed knowledge based on emotion or feeling, not reality. Secondly, coherence in presenting the truth. For instance, how does African knowledge fit into their truth claims or editorial policy and reviewers? Could African researchers' papers contradict editorial guidelines regarding focus and another principle? Put differently, could there be instances where editorial policies are designed to gatekeep or disqualify papers coming from the shores of Africa? Regardless, Jensen's explanation made a case for coherently proving the existence and occurrence of discrimination. Finally, the consensus is to bring people to believe in your truth. This is ideal for making truth universal. In other words, the truth cannot only be true in Africa but should be proven or replicated elsewhere. Otherwise, it will remain African truth only. Therefore, the claims and knowledge of possible and unconscious bias in the editorial review process can be evaluated based on justification, presentation, and acceptance.

Concluding thoughts

The market drives managerialism and capitalist approaches to knowledge production by academics working in higher education institutions. The marketisation and commercialisation of knowledge in higher education have become popular simply due to competition for funding. The pressure to publish and seek funding for research continues to create unethical practices in the search for funds. It compromises the quality of the knowledge produced by African higher education scholars over the years. Because managers oversee higher education management and governance, knowledge is seen as a countable commodity, and reward is based on the number of publications and not necessarily the quality. This is the reason for the proliferation of predatory journals and the promotion of academics inexperienced in sound research. This chapter is critical of the market approach to knowledge production in higher education research and encourages reform for research production in general. It argues and seeks answers to questions such as: what knowledge matters? Who controls knowledge in higher education research? Instead of being controlled, how can Africa facilitate North-South collaboration in knowledge sharing?

References

Ahmeti, K. (2021). Genealogy and its relation to power in the postmodern philosophy of Michel Foucault. *Trakya University Journal of Social Science*, *23*(2), 735–746.

Alajoutsijärvi, K., Alon, I., & Pinheiro, R. (2021). The marketization of higher education: Antecedents, processes, and outcomes. In *The marketisation of higher education* (pp. 17–45). Cham, Switzerland: Palgrave Macmillan.

Alemu, S. K. (2019). African higher education and the Bologna process. *European Journal of Higher Education*, *9*(1), 118–132.

Besley, T. A., & Peters, M. A. (2009). Neoliberalism, performance and the assessment of educational research quality: Comparing United Kingdom, Australia & New Zealand. In *Assessing the quality of educational research in higher education* (pp. 25–48). Leiden, The Netherlands: Brill Sense.

Bessant, S. E., Robinson, Z. P., & Ormerod, R. M. (2015). Neoliberalism, new public management and the sustainable development agenda of higher education: History, contradictions and synergies. *Environmental Education Research*, 21(3), 417–432.

Connell, R. (2017). Southern theory and world universities. *Higher Education Research & Development*, *36*(1), 4–15.

Contu, A. (2020). Answering the crisis with intellectual activism: Making a difference as business schools scholars. *Human Relations*, *73*(5), 737–757.

Croucher, G., Lacy, W.B. The emergence of academic capitalism and university neoliberalism: perspectives of Australian higher education leadership. High Educ 83, 279–295 (2022). https://doi.org/10.1007/s10734-020-00655-7

Da Silva, J. A. T., & Kimotho, S. G. (2022). Signs of divisiveness, discrimination and stigmatization caused by Jeffrey Beall's "predatory" open-access publishing blacklists and philosophy. *The Journal of Academic Librarianship*, *48*(3), 102418.

de Guevara, B. B., & Kostić, R. (2017). Knowledge production in/about conflict and intervention: Finding 'facts', telling 'truth'. *Journal of Intervention and Statebuilding*, *11*(1), 1–20.

del Cerro Santamaría, G. (2019). A critique of neoliberalism in higher education. In *Oxford research encyclopedia of education*. https://oxfordre.com/education/display/10.1093/acrefore/9780190264093.001.0001/acrefore-9780190264093-e-992;jsessionid=9A76A68AE45073D63D3DE30A02CCED7E

del Cerro Santamaria, G. (2020). Challenges and drawbacks in the marketization of higher education within neoliberalism. *Rev. Eur. Stud*, *12*, 22.

del Cerro Santamaría, G. (2021). Neoliberal higher education and its discontents. John D. Branch, Bryan Christiansen (Eds.). In *The marketisation of higher education: Concepts, cases, and criticisms* (pp. 423–453). Cham, Switzerland: Springer International Publishing.

Dougherty, K. J., & Natow, R. S. (2020). Performance-based funding for higher education: How well does neoliberal theory capture neoliberal practice? *Higher Education, 80*, 457–478.

Erickson, M., Hanna, P., & Walker, C. (2021). The UK higher education senior management survey: A statactivist response to managerialist governance. *Studies in Higher Education, 46*(11), 2134–2151.

Feldman, Z., & Sandoval, M. (2018). Metric power and the academic self: Neoliberalism, knowledge and resistance in the British University. *TripleC: Communication, Capitalism & Critique*, *16*(1), 214–233.

Gojkov, G. (2012). Postmodern pedagogy. *Educaţia Plus*, *8*(2), 8–43.

Günther, S. (2016). Ibn Rushd and Thomas Aquinas on education. In *The heritage of Arabo-Islamic learning* (pp. 250–283). Leiden, The Netherlands: Brill.

Hanesworth, C. (2017). Neoliberal influences on American higher education and the consequences for social work programmes. *Critical and Radical Social Work*, *5*(1), 41–57.

Higgs, P. (2017). Teaching African philosophy and a postmodern dis-position. In *The Palgrave handbook of African philosophy* (pp. 765–778). New York, NY: Palgrave Macmillan.

Ivanova Kremakova, M. (2016). The "New Spirit of Academic Capitalism": Can scientists create generative critique from within? *Teorie Vědy/Theory of Science*, 38(1), 27–51.

Jackson, P. A. (2017). The neoliberal university and global immobilities of theory. In *Area studies at the crossroads* (pp. 27–44). New York, NY: Palgrave Macmillan.

Klikauer, T., & Link, C. (2021). The idea of the University–A review essay. *Australian Universities' Review*, *63*(1), 65–75.

Luyaluka, K. L. (2016). An essay on naturalized epistemology of African indigenous knowledge. *Journal of Black Studies*, *47*(6), 497–523.

Lynch, K. (2010). Carelessness: A hidden doxa of higher education. *Arts and Humanities in Higher Education*, *9*(1), 54–67.

Lynch, K. (2012). On the market: Neoliberalism and new managerialism in Irish education (Working Paper). Dublin: School of Social Justice, University College Dublin.

Lynch, K. (2014). New managerialism, neoliberalism and ranking. *Ethics in Science and Environmental Politics*, *13*(2), 141–153.

Lynch, K. (2015). Control by numbers: New managerialism and ranking in higher education. *Critical Studies in Education*, *56*(2), 190–207.

Lynch, K. (2017). New managerialism in education: The organizational form of neoliberalism. In A. Abraham-Hamanoiel, D. Freedman, G. Khiabany, K. Nash (eds.), *Liberalism in neoliberal times: Dimensions, contradictions, limits*. Goldsmiths.

McCann, L., Granter, E., Hyde, P., & Aroles, J. (2020). 'Upon the gears and upon the wheels': Terror convergence and total administration in the neoliberal university. *Management Learning*, *51*(4), 431–451.

McClure, K. R. (2016). Building the innovative and entrepreneurial university: An institutional case study of administrative academic capitalism. *The Journal of Higher Education*, *87*(4), 516–543.

Medie, P. A., & Kang, A. J. (2018). Power, knowledge and the politics of gender in the Global South. *European Journal of Politics and Gender*, *1*(1–2), 37–54.

Mirzamohammadi, M. H., & Mohammadi, H. (2017). Neoliberalism and higher education in Iran: A critical perspective. *Problems of Education in the 21st Century*, *75*(5), 467.

Narkunas, J. P. (2020). Trickle-down managerialism: Accountable faculty in the financialized university of managers. *AAUP Journal of Academic Freedom, 11*(2020), 1–21.

Ngara, C. (2007). African Ways of knowing and pedagogy revisited. *Journal of Contemporary Issues in Education, 2*(2), 7–20.

Nureev, R., Volchik, V., & Strielkowski, W. (2020). Neoliberal reforms in higher education and the import of institutions. *Social Sciences, 9*(5), 79.

Parashar, S., & Schulz, M. (2021). Colonial legacies, post-colonial 'selfhood and the (un) doing of Africa. *Third World Quarterly, 42*(5), 867–881.

Patel, S. (2022). Decolonial research methods: Resisting coloniality in Academic Knowledge Production (Webinar 5). https://eprints.ncrm.ac.uk/id/eprint/4559/1/Sujata_Patel_Decolonial_Research_Methods.pdf

Peters, M. A. (2020). Digital socialism or knowledge capitalism? *Educational Philosophy and Theory, 52*(1), 1–10.

Raaper, R. (2016). Academic perceptions of higher education assessment processes in neoliberal academia. *Critical Studies in Education, 57*(2), 175–190.

Rowlands, J. (2017). The theoretical and policy context. In *Academic governance in the contemporary university* (pp. 91–110). Singapore: Springer.

Rudolph, S., Sriprakash, A., & Gerrard, J. (2018). Knowledge and racial violence: The shine and shadow of 'powerful knowledge'. *Ethics and Education, 13*(1), 22–38.

Sigahi, T. F. A. C., & Saltorato, P. (2020). Academic capitalism: Distinguishing without disjoining through classification schemes. *Higher Education, 80*(1), 95–117.

Tight, M. (2019). The neoliberal turn in higher education. *Higher Education Quarterly, 73*(3), 273–284.

Tripathi, S. (2021). International relations and the 'Global South': From epistemic hierarchies to dialogic encounters. *Third World Quarterly, 42*(9), 2039–2054.

Vican, S., Friedman, A., & Andreasen, R. (2020). Metrics, money, and managerialism: Faculty experiences of competing logics in higher education. *The Journal of Higher Education, 91*(1), 139–164.

Watson, J. (2020). *Knowledge erosion and degradation: a single case-study of knowledge risks and barriers in a multi-business organization* (Doctoral dissertation, Victoria University): Melbourne, Australia.

Wood, F. (2010). Occult innovations in higher education: Corporate magic and the mysteries of managerialism. *Prometheus, 28*(3), 227–244.

Zajda, J. (Ed.). (2020). *Globalization, ideology and neo-liberal higher education reforms.* Dordrecht, the Netherlands: Springer.

5 Epistemology realism and racism in higher education

Introduction

The realism knowledge (Lehrer, 2018) of racism can be useful in challenging different ideologies about societal issues, including in higher education settings. Realism frames knowledge about the world from our senses and how we perceive reality. Our perception of the world also influences our understanding of racism. Race and racial discrimination in higher education continue to threaten African society and its knowledge space. However, there seem to be diverging views concerning the reality of racism. Diverse perspectives about the Black race and cultural identities have emerged as a result of efforts to demonstrate the existence of racism knowledge in higher education (Feagin & Ducey, 2018; McGee, 2021; Simpson & Yinger, 2013). Different ideas, grounded in personal experiences or realities, have been utilised to frame racial discourse.

Demonstrating the existence of racism in Africa is gaining attention from Black groups and societies worldwide (Marable, 2015). African minds have always been portrayed in Western writings as superstitious. As a result, Africans will constantly disagree with Western scientific and philosophical rationality (Asante, 2017), including the Western outlook on the entire debate on race and racism. In the context of Africa, epistemology has labelled African knowledge as nonsensical for the past 500 years. Contrarily, Africans have successfully navigated the world around them by applying their accustomed techniques. Therefore, racism is multidimensional. Analysing race in higher education can provide a perspective for a focused discussion on the topic. Race has several facets, whether explicitly discussed in the context of higher education or broadly. Global discriminatory experiences could be employed to analyse race since racial coexistence is unavoidable. Moreover, racism can occur among Black communities worldwide due to differences in skin tones. In such instances, conflicts may arise between those of a darker shade and those with a lighter skin complexion. Finally, moving to a predominantly Caucasian country can lead to racial discrimination against non-Caucasian Africans.

Racial conflict may arise in higher education, for instance, when students from various origins share and produce knowledge. Multiracial student coexistence adds a new layer of complication to racism. Additionally, resolving racial conflicts in higher education institutions depends on policies that limit their occurrence and

DOI: 10.4324/9781003364740-6

the nation's history. The literature suggests supporting high-risk students to lessen the impact of racism in tertiary institutions (Harper, 2012). However, diverse and multicultural societies such as the United States have demonstrated inadequacies in addressing identity formation, critical race theory, and experiences of racial discrimination in higher education (Hiraldo, 2010). The movement of racial minorities studying abroad to end racism on campuses also draws attention to the negative connotation of 'whiteness' in higher education (Cabrera, 2014). Owing to most African nations' history, race, and identity remain challenging topics, particularly in countries such as South Africa (Motloung, 2019). Irrespective of its origin, the primary concern is to demonstrate the existence of racism in higher education.

Epistemic racism in higher education

Demonstrating a comprehensive understanding of racism and its manifestations in various circumstances is a formidable task. For centuries, colonial and neo-colonial epistemic violence has discarded, distorted, and obscured the concept of African rationality, as evidenced in countless Western scholarly publications. Darwinian evolutionism, Lévy-Bruhl's grand dichotomy, and Hegel's philosophy of world history have inevitably defined the African mind in binary terms contrary to the Western mind and the rest of mankind (Asante, 2017). One of the critical issues is selecting an appropriate approach to ethnic and racial identity racism on campus and whether resistance among racial minorities is morally valid (Rivas-Drake et al., 2022). Racism can manifest as violent outbursts towards those considered inferior due to their skin colour. Anger has become an indicator of racist behaviour since ethnic minorities suffer the perpetrators' wrath during racist conflicts.

Geographic location, i.e. Africa or elsewhere, seems to be a determinant in the responses to verbal and physical attacks towards ethnic minorities. Compared to the West, limited protection for ethnic minorities is provided in Africa despite anti-discriminatory laws enforced differently on the continent. African authorities are not likely to intervene in racial conflicts prior to violent attacks in institutions or communities. The slow response to incidences of racism is mainly political and can engender other forms of racism. Africans are often accused of conjuring instances of racism without justification. If that is the case, it will prove an intriguing quest to determine factors which shape Africans' mindsets. It could be suggested that history and ongoing discrimination pre- and post-colonisation influenced African's views on colour and their reactions to race-related issues. Black students' minds may have been shaped by stories and experiences of racist incidents in their communities. Consequently, students with this background are prone to respond strongly to deeds judging their knowledge based on their skin colour.

Generating knowledge on racism as a multifaceted concept and the reactions thereof is essential. Research reports a high prevalence of racial discrimination, particularly for mixed races of African descent, within and outside Africa, especially for mixed Races of African-Americans (Ford & Airhihenbuwa, 2010). Cultural racism among people of African-Caribbean descent has been investigated to understand the evolution of bio-racism to cultural racism (Dominelli, 2017). Similarly, research has demonstrated the harmful effects of racism, such as the failure to recognise racist views towards African and African American students (Taylor, 2019). In other words, existing literature prevents the struggle against racism since the movement critiques White domination or power.

On the other hand, the literature rarely analyses racism among Africans. Thus, two powerful forces are at play in Africa. First, former colonisers' control of knowledge and prejudice against Africans is a reality on the continent. Secondly, Black-on-Black racism and rejection of African knowledge for Western knowledge have been reported.

Though it is difficult to prove, students' reactions to poor treatment can be used to determine whether racism affects knowledge. It is incredibly challenging to prove that black students are being continuously failed due to their skin colour or that they do not experience unfair treatment in classrooms or from their peers. Methods of conveying knowledge in multicultural classrooms perpetuate present hierarchies of various knowledge types. Sustaining the status quo often triggers unpleasant responses from students in possible racist incidents in classes. Awareness of these dynamics is primordial in delicate settings. The belief that Africans cannot develop superior knowledge about complex subjects has led to discrimination against students whose participation and performance are deemed worthless. Interaction between people of different races often leads to instances of discrimination, and differentiating between the perception of widespread prejudice and actual occurrences might be a formidable task. Prejudice in the classroom might be racial discrimination against students themselves rather than against their knowledge. Though similar to those in other contexts, dissecting the topic of race in a higher education context is distinct in differentiating perceived racism from actual, demonstrable instances of racism. As a result, racial issues are profoundly problematic and multidimensional.

The following section details the multidimensionality of racial discrimination in general and in higher education settings in particular. Different lenses are employed to analyse the problematic topics of Race, racial discrimination and epistemologies. The section calls attention to the causal factors of racism rather than focusing on the occurrences and circumstances. An attempt is also made to emphasise the need to create innovative teaching and learning methods in multicultural countries and classrooms.

Multidimensionality of racism

Undermining African knowledge starts with the myth that members of the Black race are less intelligent. Intellectual slavery is implied in the argument by Africans with citizenship in Western nations that Western knowledge is superior to other types of knowledge. People of African descent seem to experience self-doubt, identity crises, and benchmarking against what they perceive to be a 'better race', i.e. the White race, due to this form of racism. Historically White institutions and elite universities in North America, Western Europe, and Africa are still designed to perpetuate this position regarding admissions and university governance.

Explaining racism's existence and its clear manifestation in mixed communities is always a contentious affair (Quaye, Harper, & Pendakur, 2019). Africa's higher education system aims to produce intellectuals to address the continent's mounting issues. For instance, selecting teachers in higher education should be subject to the basic requirement of suitable teaching methods and acceptable research records. Similarly, applicants to higher education institutions ought to be accepted regardless of race or origin. In theory, higher education policy ought to promote innovation for race and ethnicity to accommodate diversity (Stewart, 2019). Recruitment processes that enable the eradication or limitation of racial discrimination in institutions should be thoroughly researched and implemented. Alas, a stumbling block presents itself in implementing those policies promoting acceptance and diversity, which could provide a different perspective on racism in higher education.

There is a multitude of viewpoints concerning racism, which makes it one of the most challenging problems to address in African higher education. Every theorist develops their epistemology by assuming the inevitable reality of racism. African epistemology is based on a comprehensive perspective that acknowledges the interconnectedness of all realities (Asante, 2017). The historical narrative of racism's occurrence in a different setting is linked to African theories. Racism occurrences in higher education are observed during knowledge dissemination and assessment processes.

Instances and occurrences of racism manifested as xenophobia, afrophobia, ethnic violence, and other behaviours have been reported in several African universities worldwide (Dowd & Bensimon, 2015). These occurrences lead to violence and affect student-academic relations in post-colonial Africa. Racial tensions frequently generate fear and require collective resistance (Kubota, 2020). However, collective resistance terminologies have been misinterpreted as violent or retaliatory movements. Lacklustre responses to instances of racism reported to the government or higher education leadership often trigger violent reactions. Therefore, the victims' only means of demonstrating the existence of racism is to cause disruptions in the classroom or, occasionally, to cause harm to perpetrators or supporters of racist incidents on their campuses or in their communities. Open discussions and sincere efforts to address racial conflicts can help to lessen the distress caused by racial discrimination. Africans may continually want to demonstrate the truth of racism, but regrettably not in a non-violent manner, if it is suggested that racism is an African mental construct.

Collective resistance to challenge and influence higher education policies is needed. Having a shared knowledge of the issue is therefore necessary. This is challenging with racism since, in most cases, the root causes are not addressed. Due to the exclusion of Black people and other racial and ethnic minorities, Afrocentrism and nationalism, which have developed to reproduce another type of racism and discrimination, have historically been opposed in several ways (Clarke, 2017). Critical race theory is used in many fields to problematise racism and the hegemony of the White race (see Cabrera, 2019). In academic writing, conferences, and policy meetings, the approach to demonstrating racism's reality remains challenging. Some academics suggest that racism is either overemphasised or a myth in the twenty-first century. Therefore, it is not widely known that racism or anti-black racism may exist in every African institution (Allen, McLewis, Jones, & Harris, 2018). Thus, collective resistance should focus on academic issues and be directed at the root causes of racism on campuses rather than analysing the victims' thoughts about their experiences.

Some contend that not every instance of racism can be connected to historical events. Yet, it is challenging for higher education research to prove that racism is unrelated to history. The reality of racism in higher education, from a historical viewpoint, must first be presented. Many racist incidents were linked to the personal experiences of mistreated individuals. Research by Changamire et al. (2022) brought racism-related issues to light from the perspective of sub-Saharan African students studying in the United States concerning their institutions' stated priorities. The difficulty lies in how people from different backgrounds view, express, and write about racism in higher education systems. Thus, the idea of the power of minds explains the epistemological realism of racism in higher education. A body of research has attempted to demonstrate that racism is either a mental construct or only relevant in the contexts of Brazil, Canada, South Africa, the United Kingdom, and the United States (Wong, Elmorally, Copsey-Blake, Highwood, & Singarayer, 2021). The discussion ought to centre on cultivating an anti-racist and inclusive institutional culture (Tate & Bagguley, 2017). Racism can be demonstrated in policies as existing but with the purpose of being rectified by using lived experiences and perceptions of the subject matter. In other words, it can be considered that racism does not exist and that incidents deemed racial could be caused by factors excluding race. Some subject matter experts affirm that politics or policy may fuel anti-racism movements though no tangible evidence has been provided to validate the existence of racism.

Interpreting racism

Research has attempted to prove the significance of racism in African higher education (Ogbar, 2019). This covers the racialisation of Africans in Europe between 1870 and 2000, including non-Africans (MacMaster, 2017) and African refugee students in Ireland, among others (Sheridan, 2021). However, Racism cannot be proven in every circumstance. On the other hand, many believe that racist historical events should be left in the past, as continuously referring to them encourages people to attribute every conflict in higher education classrooms and workplaces to racism (Glasgow, Haslanger, Jeffers, & Spencer, 2019). The realist school of thought emphasises experience and the idea that current circumstances can jog memories of past events. Racism is a growing concern as it spreads worldwide through politics and policy. Immigration laws purposefully discriminate against or exclude some African citizens based on factors including their skin colour and educational background, among other factors. Policy provisions are often made to make African students less welcome than immigrants from Europe.

A counterargument contends that it is fallacious to assume that racism is the causal factor of Africans and other ethnic minorities' exclusion and limitation. A diverging viewpoint holds that as long as the exclusion of minorities is morally acceptable, it is not racism. Such conversations highlight the need for Afro-centric epistemologies and perspectives to address racism-related concerns. A potential solution could be developing and promoting a curriculum that advocates African unity. This could mark the start of the learning and knowledge-producing process that will teach Africans to cherish their heritage, culture, and language.

Regarding the realism of knowing through experience and the mind regarding racial concerns, Peirce, Cohen, and Dewey (2017) argue for the imperativeness of elucidating our knowledge to distinguish between perception and reality. A thorough investigation is needed into the claim that people's minds merely project racism in various higher education contexts (Keval, 2018). Researching racism and prejudice is inconceivable without analysing them through the lens of different past experiences. Knowledge of historical events assists in moulding and shaping minds and their beliefs concerning future treatment. Knowledge of the past and the workings of African students' minds in various circumstances may be of considerable assistance. Ignorance of historical events may negatively affect the youth's renderings regarding issues of racism, which may lead to misconceptions.

The narrow conceptualisation of racism in higher education (Cazenave, 2015) can be daunting without proper expertise in interpreting racial discrimination. Thus, the dimension of racism based on ethnicity, gender and nationality signal multiple types of discrimination (Vieten, 2011) and the various methodologies in their analysis. For instance, scholars frequently discuss women and girls' lack of equal access to higher education, as demonstrated in the cases of South Africa (Walker, 2018), Kenya (Odhiambo, 2011), and other sub-Saharan Africa countries (Johnson, 2014). Though cultures and systems resistant to change hinder authentication processes, such research validates personal claims of gender-based discrimination.

African academics' discussions on racism are often heavily based on observational analysis, which may produce inaccurate findings. Consequently, higher education curricula are scrutinised to determine whether historical events related to racism are highlighted or excluded (Ladson-Billings, 2019). It might be necessary to include material on racial conflicts in higher education curricula and instruction. Utilising persuasive approaches might be required to convey the complexity of racism for Africans in particular. According to Asante (2017), African epistemology rejects all types of dualism, with the major dualism being the relationship between the subject and the object of the investigation. Africans interact with the subject of the study rather than distance themselves. Essentially, they consider mixing various tools or human experiences to be the most efficient approach to learning about any topic. Therefore, the core of the African holistic approach is epistemological discourse or interdisciplinarity.

The next section provides a guideline for identifying a research approach to study racism in higher education. The explanation focuses on the need to analyse both the literature and theories relevant to it. A research problem must be formulated from the literature of history and sociology to contribute to knowledge about racism in Africa. In other words, a researcher should understand this field of study through an extensive review of past events and societal trends. The process further includes understanding what different authors have written about the topic. It may also include ignored or incoherent issues that require further rethinking. Filling this gap is essential, as expert knowledge in the current race issues is primordial. However, using a quantitative design to study race might be inadequate since the phenomenon or issue of racism cannot always be measured or manipulated statistically to arrive at a conclusion. The issue of race can often be explored through histography or the ethnographic analysis of the origin of racism in higher education and Africa.

Realism and human reality of racism

According to realists, logical determination of knowledge components will instruct the implementation of anti-racism policies (Diffey & Mignone, 2017). This entails grappling with the real-world issues of racism, such as those that arise in higher education. It also advises using an alternative methodological paradigm to know racism's truth on various levels without ignoring racism's historical human experience. This argument contends that historical occurrences should serve as a foundation for the study racism (Stewart, 2017). Rational thought is required to demonstrate the actuality of racism rather than relying on prior personal knowledge or experiences, which makes us subjective in our analysis.

Contentment is paramount for the growth of any society. As such, a sense of belonging should be maintained in a university environment regardless of race, which involves a number of variables. Otherwise stated, academic and logical discussions of racism in the classroom might increase the likelihood of students or academics thriving and realising their full potential. The method used to frame and discuss racism in multicultural or multiracial classrooms is another concern which may impact research, teaching, and learning approaches. Due to the complexity of racism (Harbin, Thurber, & Bandy, 2019), various ways have been devised to understand it. The difficulty in implementing anti-racism policies in higher education stems from the failure to reach a consensus on the logic and approach for teaching its reality. Furthermore, it is construed as dominance or exploitation in social sciences to apply realist ideology to racism (Mahmud & Zaman, 2022). Higher education might be considered distinctive and offer new perspectives on racism. Since contributing factors to racism in various African countries and universities vary, it may be inaccurate to analyse racism using the same methodologies. The question remains to determine how organisations can ensure their culture is shaped by plausible policies that explicitly state their stand regarding racism.

In higher education, selecting a historiographic analysis of racism from which broad conclusions may be reached is comprehensive. Such an approach offers the assistance required to forge the connection between the past and present. Using historical data to draw general conclusions about occurrences enables the researcher to present impartial theories about the issue. Indeed, the most accurate historical views come from the collection of anthropological and archaeological evidence. This kind of investigation may be required if a researcher is interested in learning about students' opinions towards race in multicultural societies like South Africa, Liberia, Uganda, and Tanzania. Before conducting historical research, it is crucial to understand the history of culture and significant post-independence transformations. In this context, views can be compared with the history that students have been socialised into from birth rather than experiences or perceptions. This approach would enable researchers to determine the type of information needed to determine whether students' experiences reflect their fear of being subjected to racism or reveal other causal factors.

On the other hand, research problems in relation to race or ethnic race require the researcher to explain the nature of the link between history and personal experiences. Explaining an existing relationship is essential in determining whether one or more societal matters have influenced viewpoints on racism or whether racism was a consequence of those issues. Table 5.1 provides knowledge trends on the different dimensions of racism in African higher education.

Epistemological racism occurs in higher education classrooms when teaching and learning methods marginalise people of colour (Lachaud, 2020). Similarly, research by Talmacs (2020) presents Chinese university students' negative perception of Africa and Africans through racially discriminatory language adopted from Hollywood movies. The negative depiction of people of colour in various films could be attributed to the media narratives on Africa broadcasted in their societies. Messages derived from such films could also influence individuals' perceptions of people of African descent. Hence, the assimilation of a negative, media-derived narrative showcases the power of media to propagate racism.

Based on their viewing experiences, Chinese students' minds, especially those with little knowledge of Africa and its people, have been captured. It is worth considering how to reverse these harmful teachings. A curriculum teaching Africa's real history and the meaning of being human, regardless of race, is one way to address this. In the absence of such a step, these students will struggle to co-exist peacefully and in harmony with African international students. The urgency of this relearning process is further accentuated by the racism directed at African students in China's higher education.

Similarly, Arafah and Pattu (2022) have revealed from a poem written by Langston Hughes how people of colour are portrayed with different derogative names, such as enslaved people, among other names. Name-calling is one of the means employed to disturb the mind of Black people, for instance. Such discrimination significantly affects black students' confidence in countries where they are constantly exposed to inadequate treatment. In higher education institutions, name-calling is often the reason for black students resorting to violence after lengthy verbal and emotional abuse. There have been cases of rustication of black students for reacting violently, and implemented policies further aggravate already delicate situations. Without justifying violent reactions, it is crucial to evaluate instances of racial violence beyond the obvious.

Cases of violence due to racial discrimination are expected in the United States. Joseph, Peterson, Gordon, and Kamarck (2021) explain how racial bias may have underscored prevalent cases of psychological problems and negative emotions among African-Americans. In the case of students, negative emotions can be addressed once reported, though university management's questionable inquiry into instances of racism may discourage students from reporting such incidents. Unreported racist conflicts or incidences can develop into anger, which may force victims to resort

Table 5.1 Different dimensions of racism in higher education

Dimensions	*Source*	*Arguments in the knowledge*	*Knowledge analysis*	*Citation*
Racism against Africans in China	Journal	It is argued that these responses must be understood within the context of the trust that students have in the mediated messages they encounter, the Chinese leadership, the hearsay of social networks, and the film industry standards established by Hollywood, all of which precondition Chinese students' understandings of Africa and Africans.	This paper analyses responses from Chinese university students to China's most successful blockbuster to date, *Wolf Warrior 2*. Responses revealed racialised language objectifying the black African Other and affirming existing scepticisms towards Sino-African relations.	Talmacs (2020)
Racial discrimination of Black Africans in Langston Hughes's Poems	Journal	The result showed that Langston Hughes reflects the phenomena of racial discrimination through his poems. Terms such as slave, victim, nigger, torture, darker brother, and not beautiful are employed in his poems. Langston Hughes, in his poetry, affirms that Blacks are poorly treated: they eat in the kitchen, they are not beautiful, and children's happiness is tarnished. He concludes that racial discrimination is inhumane.	This study aims to describe the racial discrimination from White people against former black slaves by analysing Langston Hughes' poems, *I, Too, To the Black Beloved, The White Ones, and My Beloved.* Presentation of racial discrimination can be seen from the act of prejudice, insulting, words used, and the suppression of black people.	Arafah & Pattu (2022)
Racial discrimination against Black Africans and people of African descent outside of Africa	Journal	Racial discrimination is a common experience for African Americans, but no research has examined how discrimination reported in daily-life moments influences concurrent negative emotions and psychosocial resources. These results are the first to demonstrate that personal and vicarious racial discrimination is associated with negative emotions and lower coping resources in daily life moments and that contextual factors modify these associations.	Results refine our understanding of the immediate sequelae of discrimination in daily life and point to possible targets for brief ecological interventions.	Joseph, Peterson, Gordon, & Kamarck (2021)

(Continued)

Table 5.1 (Continued)

Dimensions	*Source*	*Arguments in the knowledge*	*Knowledge analysis*	*Citation*
Racial discrimination among Black Africans in Norway	Journal	With a sample of 162 African first- and second-generation immigrants living in Norway, the present research investigated the relationship between both types of discrimination, acculturation and psychological well-being. It did so by focusing on discrimination based on one's skin tone, a kind of discrimination Africans can experience from White and African individuals. Results showed that skin-tone discrimination by Whites was associated with a lower host culture orientation. By contrast, African skin-tone discrimination was associated with a more downwards heritage culture orientation. Mediation analysis showed that the positive relationship between skin-tone discrimination by Whites and Africans with life satisfaction was mediated by a lower host and heritage culture orientation, respectively.	It is well-established that experiences of discrimination influence immigrants' acculturation. Yet, whereas a large body of research has demonstrated the role of discrimination by members of the dominant societal group, surprisingly, little is known about how being discriminated by members of one's own group relates to how immigrants acculturate.	Kunst & Phillibert (2018)

to violence. Taking another example of racism against Black Africans in Norway, Kunst and Phillibert (2018) conclude that Africans, including students, experience racism everywhere owing to their skin colour. Yet, difficulty in providing evidence beyond experiences of racism prevents them from obtaining justice.

Conclusion

Throughout research about epistemological racism in higher education, little emphasis is placed on the reason for the difficulty in proving and addressing racism. Existing studies inadequately discuss how racism can be resolved through challenging teaching and learning approaches in African higher education. Thus, there is a problem with the way racism is viewed and discussed in higher education and academic literature.

One crucial question is why it is difficult for Africa to share similar knowledge about issues of racism. Research on the topic engages Whites discriminating against Blacks but omits that racial discrimination can occur against Black people due to neo-colonial ideologies about their identity. The idea is to frame racism as a migration problem. Given this premise, it is not clear whether the literature is theorising racism as a perception issue or as a human-caused problem. The weakness of most research in this field is the lack of solid engagement in literature and history. Studies are crowded with different thoughts on how racism is perceived in a particular context, why racism is a complex topic and why anti-racism policy is problematic in many African countries. Research should instead focus on exploring the interdisciplinary subjects of sociology and history to reframe arguments about racism in African higher education. Though there is evidence that the existing literature is critical of how racism is being discussed by non-African researchers, the best approach could be to evaluate different thoughts from the grassroots.

Numerous organisations are battling racism in Africa and globally, beyond the higher education sector. Therefore, the first step is to criticise texts that perpetuate racism. These organisations require knowledge of the reality of racism and the best way to address the problem and its root causes. The crucial topic of racism's fundamental causal factors in Africa is either not discussed in the literature or is not well stated. Additionally, African scholars' different opinions inhibit any real progress. It is critical to discover how the varied knowledge of racism in higher education affects Africa's development or progress? In other words, it is essential to determine how the different knowledge domains and their dissemination affect peace on higher education campuses and the continent. We can, therefore, suggest that writings and research on racism can be redirected to influence thinking that will cause peace and development among African nations instead of discrimination and high cases of xenophobia.

Many lessons are to be learned from past research on racism. In summary, the findings indicate that top-down approaches to solving problems of racism and/or discrimination often fail to bring about desirable changes in Africa. Likewise, political failure has led to undermining factors igniting all forms of racism on university campuses. Many outputs have contributed to the transdisciplinary nature of racism in higher education. However, research still needs to show that racism is not an illusion and not a trick of the mind for African minorities who have experienced it on their continent and elsewhere.

African students' encounters with racism both within and outside of their continent are trending upwards, according to statistical data. Through statistics and experiences of victims, it is feasible to conclude that racism exists on university campuses around the world. Some speculate that this issue is especially problematic for international students, particularly African students studying in the United States. However, opposing views emphasise the importance of justifying knowledge rationally and objectively. Similarly, the problematisation of race that makes a case for radical change and critical race theory in curriculum transformation is another issue that creates problems in research. It will be interesting to understand the policy context of race and curriculum transformation in countries where cases of racism are affecting many African students. Rejecting change through curriculum transformation is evidence of racism or confirmation of its problematic nature.

References

Asante, M. K. (2017). The philosophy of Afrocentricity. Adeshina Afolayan, Toyin Falola (Eds.). In *The Palgrave handbook of African philosophy* (pp. 231–244). Palgrave Macmillan New York.

Allen, W. R., McLewis, C., Jones, C., & Harris, D. (2018). From Bakke to Fisher: African American Students in US higher education over forty years. *RSF: The Russell Sage Foundation Journal of the Social Sciences*, *4*(6), 41–72.

Arafah, B., & Pattu, A. (2022). Racial discrimination experienced by black people as reflected in Langston Hughes's poems. *Journal of Language Teaching and Research*, *13*(2), 350–356.

Cabrera, N. L. (2014). Exposing whiteness in higher education: White male college students minimizing racism, claiming victimization, and recreating white supremacy. *Race Ethnicity and Education*, *17*(1), 30–55.

Cabrera, N. L. (2019). *white guys on campus: Racism, white immunity, and the myth of "post-racial" higher education*. New Brunswick, NJ: Rutgers University Press.

Cazenave, N. A. (2015). *Conceptualizing racism: Breaking the chains of racially accommodative language*. Lanham, Maryland, United States: Rowman & Littlefield.

Changamire, N., Mwangi, C. A. G., & Mosselson, J. (2022). International students and The neoliberal marketplace of higher education: The lived experiences of graduate students from sub-Saharan countries in Africa of a US university's internationalization policy. *Higher Education*, *84*(3), 505–521.

Clarke, S. (2017). *Social theory, psychoanalysis and racism*. London, England: Bloomsbury Publishing.

Diffey, L., & Mignone, J. (2017). Implementing anti-racist pedagogy in health professional education: A realist review. *Health Education and Care*, *2*(1), 1–9.

Dominelli, L. (2017). *Anti-racist social work*. London, England: Bloomsbury Publishing.

Dowd, A. C., & Bensimon, E. M. (2015). *Engaging the "race question": Accountability and equity in US higher education*. New York: Teachers College Press.

Feagin, J. R., & Ducey, K. (2018). *Racist America: Roots, current realities, and future reparations*. Routledge.

Ford, C. L., & Airhihenbuwa, C. O. (2010). Critical race theory, race equity, and public health: Toward anti-racism praxis. *American Journal of Public Health, 100*(S1), S30–S35.

Glasgow, J., Haslanger, S., Jeffers, C., & Spencer, Q. (2019). *What is race?: Four philosophical views*. Oxford, England: Oxford University Press.

Harbin, M. B., Thurber, A., & Bandy, J. (2019). Teaching race, racism, and racial justice: Pedagogical principles and classroom strategies for course instructors. *Race and Pedagogy Journal: Teaching and Learning for Justice*, *4*(1), 1.

Harper, S. R. (2012). Race without racism: How higher education researchers minimize racist institutional norms. *The Review of Higher Education, 36*(1), 9–29.

Hiraldo, P. (2010). The role of critical race theory in higher education. *The Vermont Connection, 31*(1), 7.

Johnson, A. T. (2014). Performing and defying gender: An exploration of the lived experiences of women higher education administrators in sub-Saharan Africa. *Educational Management Administration & Leadership, 42*(6), 835–850.

Joseph, N. T., Peterson, L. M., Gordon, H., & Kamarck, T. W. (2021). The double burden of racial discrimination in daily-life moments: Increases in negative emotions and depletion of psychosocial resources among emerging adult African Americans. *Cultural Diversity and Ethnic Minority Psychology*, *27*(2), 234.

Keval, N. (2018). *Racist states of mind: Understanding the perversion of curiosity and concern*. Routledge: London.

Kubota, R. (2020). Confronting epistemological racism, decolonizing scholarly knowledge: Race and gender in applied linguistics. *Applied Linguistics*, *41*(5), 712–732.

Kunst, J. R., & Phillibert, E. N. (2018). Skin-tone discrimination by whites and Africans is associated with the acculturation of African immigrants in Norway. *PloS One*, *13*(12), e0209084.

Lachaud, Q. (2020). *Combatting epistemological racism: critical race participatory action research toward the promotion of faculty critical race conscience and transformative pedagogy* (Doctoral dissertation, University of Pittsburgh). ProQuest, Ann Arbor, MI.

Ladson-Billings, G. (2019). Just what is critical race theory, and what's it doing in a nice field like education? In *Race is... race isn't* (pp. 7–30). Routledge: London.

Lehrer, K. (2018). *Theory of knowledge*. New York: Routledge.

MacMaster, N. (2017). *Racism in Europe: 1870-2000*. London, England: Bloomsbury.

Mahmud, A., & Zaman, F. (2022). Critical theory in social research: A theoretical and methodological outlook. In *Principles of social research methodology* (pp. 83–99). Singapore: Springer.

Marable, M. (2015). *How capitalism underdeveloped black America: Problems in race, political economy, and society*. Chicago: Haymarket Books.

McGee, E. O. (2021). *Black, brown, bruised: How racialized STEM education stifles innovation*. Cambridge, MA: Harvard Education Press.

Motloung, S. M. (2019). *Analysis of race and racism discourse by academics in post-apartheid higher education* (Doctoral dissertation). University of Kwazulu-Natal, Durban.

Odhiambo, G. (2011). Women and higher education leadership in Kenya: A critical analysis. *Journal of Higher Education Policy and Management*, *33*(6), 667–678.

Ogbar, J. O. (2019). Black power: Radical politics and African American identity. Baltimore, MD: John Hopkins University Press.

Quaye, S. J., Harper, S. R., & Pendakur, S. L. (Eds.). (2019). *Student engagement in higher education: Theoretical perspectives and practical approaches for diverse populations*. Routledge: New York.

Rivas-Drake, D., Pinetta, B. J., Juang, L. P., & Agi, A. (2022). Ethnic-racial identity as a source of resilience and resistance in the context of racism and xenophobia. *Review of General Psychology*, *26*(3), 317–326.

Sheridan, V. (2021). Counter stories: Life experiences of refugee background mature students in higher education in Ireland. *Race Ethnicity and Education*, 1–18.

Simpson, G. E., & Yinger, J. M. (2013). *Racial and cultural minorities: An analysis of prejudice and discrimination*. New York: Springer Science & Business Media.

Stewart, D. L. (2017). Race and historiography: Advancing a critical-realist approach. *Journal of Diversity in Higher Education*, *10*(2), 149.

Stewart, D. L. (2019). Envisioning possibilities for innovations in higher education research on race and ethnicity. *Journal Committed to Social Change on Race and Ethnicity (JCSCORE)*, *5*(1), 7–32.

Talmacs, N. (2020). Africa and Africans in wolf warrior 2: Narratives of trust, patriotism and rationalized racism among Chinese university students. *Journal of Asian and African Studies*, *55*(8), 1230–1245.

Tate, S. A., & Bagguley, P. (2017). Building the anti-racist university: Next steps. *Race Ethnicity and Education*, *20*(3), 289–299.

Taylor, E. (2019). Critical race theory and interest convergence in the desegregation of higher education. In *Race is… race isn't* (pp. 181–204). New York: Routledge.

Vieten, U. M. (2011). Tackling the conceptual order of multiple discrimination: Situating different and difficult genealogies of race and ethnicity. *European Union non-discrimination law and intersectionality*. Farnham, England: Ashgate, 63–77.

Walker, M. (2018). Aspirations and equality in higher education: Gender in a South African university. *Cambridge Journal of Education*, *48*(1), 123–139.

Wong, B., Elmorally, R., Copsey-Blake, M., Highwood, E., & Singarayer, J. (2021). Is race still relevant? Student perceptions and experiences of racism in higher education. *Cambridge Journal of Education*, *51*(3), 359–375.

6 Epistemological access to corruption in higher education

Introduction

The topic of epistemologies (Nell, 2021: Nyamnjoh, 2019) in Africa requires an in-depth inquiry into pre- and-post-colonial histories. Africans' conceptualisation and validation of their knowledge is termed African epistemology (Nkulu-N'Sengha, 2005). African epistemology can be described as a theory of knowledge with a unique approach to discovering the truth concerning various issues affecting African society, including corruption in higher education (dos Santos Soare, 2019; Harnois & Gagnon, 2022). However, the pursuit of knowledge in an African context cannot be achieved without an insight into the history of knowledge in Africa. History can provide better insight into the Global North's control over knowledge systems (Connell, 2020). Therefore, Africa's liberation struggle from the effects of Euro-North American-centric intellectual and epistemological dominance (Ndlovu-Gatsheni, 2015) should be analysed further. The relevance of this issue lies in the fact that Western epistemological ways of providing solutions to societal problems including the problem of corruption have hegemonised African ways of knowing, expressly in social sciences or theories (Ndlovu-Gatsheni, 2018).

Nevertheless, Allais (2016) argues that Western philosophy cannot be rejected in problematising African epistemologies. On the other hand, O'Neal (2020) explained how colonialism influenced corruption in West Africa. Problematising African ways of knowing without going back to intellectual slavery remains crucial, although science appears reluctant to accept African-produced knowledge to solve modern problems. Western epistemology is expected and encouraged to develop an awareness of true African history and culture (Mudimbe, 2020). True African history means history told and described from an African perspective without Western influence. Instead, Western students showcase scepticism regarding knowledge coming from Africa (Erasmus, 2020). This type of scepticism, disguised fear, appears to be motivated by a curriculum that portrays African science, i.e. knowledge, as dangerous, sinister, superstitious, and unsophisticated knowledge. Thus, transformation (Aina, 2010), curriculum reconstruction (Higgs, 2016), and epistemological change (Heleta, 2016) are needed in this era of globalisation. Africans usually migrating to North America and Western Europe for knowledge acquisition. Interestingly, this migration does not

DOI: 10.4324/9781003364740-7

occur the other way around, further confirming the apparent reluctance towards African ways of learning. Western scientists think African knowledge requires the validation of modern Western epistemologies (Luyaluka, 2016). Essentially, discovering means of establishing and validating the truth about corruption using southern epistemologies and African theory of knowledge in higher education remains a pressing matter (see Mede & Schafer, 2020; National Academies of Sciences, Engineering, and Medicine, 2018).

While corruption is a global issue, its analysis in the African context transcends the prescription of its nature and solution from Western epistemologies. From an African perspective, discussions on corruption lead to the following key questions: how do we establish that Africa's corruption is embedded in the history of colonialism? How do we correct the notion that knowledge from Africa cannot be free of corruption? How would Africans use existing narratives to redress the issue of corruption in Africa's higher education? These questions point to what true African history is and those who have the power to share this knowledge globally. African epistemic freedom is necessary to challenge knowledge and knowers about issues such as corruption in higher education. Accordingly, this chapter contributes to the debate on post-colonial and decolonial views on knowledge utilising a philosophical and historical reconstruction of African history.

The following section showcases historical evidence of knowledge domination in Africa. The analysis includes various issues from the literature on epistemic violence and social movements that challenge knowledge imperialism and promote African philosophy to gain freedom in the post-colonial era. Further, the corruption issue is highlighted broadly, demonstrating its dynamic in Africa. The argument is that students are central to the objective of knowledge in higher education. However, the mismanagement of resources in African higher education institutions stems from the different movements and violence over the years. The section considers the possibility of contributing to these issues without epistemological access to Western theories while addressing global issues such as corruption.

Higher education in the post-colonial period

The purpose of higher education during colonialism was to create a small elite group that would adopt Western epistemologies and theories in higher education after independence (Agbo, 2005). Southern epistemological intellectual access and knowledge transfer are still reflected in higher education's post-colonial curricula and instruction. It is challenging to argue in literature or to African politicians that higher education should not be primarily used for financial gain. The proliferation of neoliberalism and capitalism is due to the misunderstanding of higher education goals, which has led to corruption in various organisations. The decolonisation of minds and higher education transformation are necessary and will require consensus and a non-resistance attitude towards reconfiguring the future of African knowledge systems. The future of Africa is about understanding different views of scholars on decolonisation.

On the one hand, some African scholars believe decolonisation is necessary for Africa's transformation (Ndofirepi, Mathebula, & Gwaravanda, 2020). On the other hand, others regard decolonisation as an overemphasised topic, and Africans cannot ignore Western knowledge in transforming knowledge acquisition and dissemination in higher education (Knaus, Mino, & Seroto, 2022). Finally, it means that some African authors still seek to carry on the tradition of Western epistemological access. The decolonisation of the mind is redefining higher education's goal to alter students' perspectives both during their studies and after graduation. The decolonisation of higher education could also mean educating African minds to create meaning based on their experiences. This includes the ability to evaluate the foundation of what they consider knowledge and whether it reflects its reality for the future of Africa. Thus, the question remains whether African miseducation can be reversed by simply changing the approach to knowledge design and corruption problem in higher education.

Countries such as Uganda (Mittelman, 2017), Cameroon and South Africa (laga Ramoupi & Ntongwe, 2017), Namibia (Callewaert, 2019), Anglophone and Francophone Africa-Benin and Ghana (Lassou, Hopper, Tsamenyi, & Murinde, 2019) have introduced reforms to challenge non-progressive ways of educating their population. These reforms have also faced patrimonial autocracy and sometimes led to civil conflicts because corrupt politics benefit from noneconomic reconfiguration (Young, 2004). Policies that stop change and contest the resistance to decolonisation cause challenges in solving significant corruption problems in Africa. The elites want to keep the poor poorer and wish to remain in power to control policies in higher education also. Colonial curriculum instils in African minds that Eurocentric values and Western epistemological tradition are a pathway to being globally competitive (see Heleta, 2016). Many African communities will invest heavily in foreign degrees to join the Western elite group. For personal gain, this system permits and redistributes power and wealth to African elites who also work as politicians. The elite is not likely to promote change and a revolution that weakens their ability to amass a fortune at the expense of others. The most effective strategy to ensure an egalitarian society is to educate African students about the value of higher education in their continent. Without this, conflicts may continue to occur inside and outside African institutions to stop the tyranny of African authorities unwilling to reform. African epistemology needs to question the neoliberal agenda (Hwami, 2023; Ndofirepi, 2023), respond to violence caused by corruption (Brunsdon, 2019; Ndelu, 2017) and politics of knowledge in higher education (Spivak, 2016). Usually, religion and politicians' opposition to the decolonisation agenda in higher education are the leading causes of the conflicts recorded in Africa.

Religion and corruption in higher education

Between 1990 and 2010, 47 African nations were the site of 7,200 conflict-related cases, according to Salehyan et al. (2012). Europeans implemented Islam and Christianity in Africa before their departure and promoted these religions above the local ones. The foundation of higher education during the colonial era was also based on these religions (Heleta, 2016). The knowledge acquired by early Africans was imparted by faith-based curricula taught in institutions. Still, discord surrounding African knowledge in health sciences for instance can be attributed to cultural and religious sentiments (Drame et al., 2019). The African approach to knowledge can be understood from the wisdom of oral tradition, such as folktales, myths about creation, and proverbs; the avenues of seeking the truth in politics, religions, and social institutions; the work of traditional healers; ways of mediating family disputes, finding guilty parties in conflicts and other social conflicts (Nkulu-N'Sengha, 2005). However, science expects Africans to prove these approaches before they are recognised as knowledge for solving modern problems.

Western religions also shape politics and society. Prior to the arrival of colonists in Africa, knowledge of African deities was in no way a source of strife. It is intriguing to learn that the philosophy of religion may also influences higher education studies and Africans' conception of wealth accumulation (Sanneh, 2015; Terblanché-Greeff, 2019). Today, various African higher education institutions are fundamentally religious (see Mohamedbhai, 2020). These religions have different conceptions of values and multiple approaches to understanding social issues and instructing students from diverse backgrounds. Alas, African students may have been presented with other curriculum perspectives that also may affect their thinking about wealth accumulation as the purpose of higher education degrees. This issue puts ethical theory in Africa into question. Ethics refers to what is morally right or wrong, which is debatable philosophically. The differences between the knowledge of ethics and morality often generate violence in Africa.

Ethics and morality of corruption

In terms of ethics and morality, one ponders the necessity of demonstrating that corruption or amassing wealth is ethically wrong in Africa. The significance of morality and ethical misconduct cannot be confirmed without considering other hypotheses and the development of religion in Africa. Some movements oppose Western ideology in several African nations and push for secularising African higher education systems (see Sklar, 2015; Smith, 2014). This implies that politics and religion will impede campus peace and may be the cause of the prevalence of corruption in African society. Decolonising knowledge and African minds by challenging moral and ethical systems fundamental to African higher education are inspired by spiritualism and morality.

The idea of ethics in Africa is founded on theological doctrine, and in South Africa, it is coined as Ubuntu (Joseph, 2018). Ubuntu means that all humanity is united by the conviction that sharing is a fundamental human value. Most African universities aspire to structure their instruction around respect-based ethics and values. However, not all African institutions fit this description. While some universities mostly want to teach love, others' curricula encourage wealth accumulation, supporting inequality or corrupt ideologies (Uslaner, 2017). This remains the situation in many African higher education institutions, contributing to the difficulty of discussing critical issues such as gender (Maistry, 2021). Higher education systems' ethical policies must be open to questioning by students.

In Africa, various ethical and value systems have their roots in religion. Though it is often disregarded, understanding the reasoning behind university-taught subjects may be an approach to deciphering these systems (Broudy, 2017). To live fulfilling personal and professional lives, students would like to comprehend the fundamentals of various institutions' value systems. Alas, they are given a variety of ethics, beliefs, and knowledge based on the curriculum creators' backgrounds without any rationale. For instance, what is the morality or justification for learning a foreign language and assessing students in a language different from their home language? The common argument (Aziz & Hashim, 2022) is that English is the global language or language of science and may give students access to career advancement opportunities. Epistemological access through English is seen as an opportunity to become part of the elite group, assimilate Western manners and mindsets, and accumulate wealth as a result. This implies that African students who lack fluency in the English language may not do well in assessments and may likely have limited opportunities. However, after studies, the benefits of acquiring the English language are an ongoing debate. There is a limit to the knowledge and morality we can impart to African students. Africans pass on their values and Western ideologies to younger Africans across generations. Younger generations are now immersed in new experiences and want minimum interference while they explore their culture and morals.

Reluctance to abandon colonial orientations

The reason behind older Africans' reluctance to abandon colonial moral codes promoting widespread corruption remains a mystery. Due to their orientation to Western values, elites regard it as a force that cannot be contested and accept it as a legitimate aspect of contemporary society. The decolonisation of minds and knowledge in Africa challenges the Western dominant power and Africans who resist change. Students in Africa question the prescription of knowledge and a value system that promotes societal corruption or one they cannot assimilate. In Ghana, Kenya, Nigeria, and South Africa, higher education should serve as a place where knowledge and moral systems can be transformed without corruption (Unterhalter et al., 2018). Yet, this is no longer the case as external forces such as politics shift higher education focus back to resource control conflicts in Nigeria (Adejumobi, 2017; Buba, 2020; Fagbayibo, 2019), problem of gender equality in Uganda (Odaga, 2019), rape culture, and corruption in South Africa (Cook, 2021; Tella, 2021). Religiosity, African epistemology and African metaphysics (Chimakonam & Ogbonnaya, 2021; Mzwazi & Shumba, 2022) should reflect a value system that can shift the attention of Africans from wealth accumulation to peace and respect for others regardless of religious views. This is a challenge as politicians and society have different opinions of higher education outcomes. It would be intriguing to uncover the truth behind Africans' wish for higher education to be more responsive to societal needs. Though such a discussion can be undertaken from both sociological and economic perspectives, the history of Africa and knowledge could prove more relevant.

The following section critiques different methods used to study corruption in higher education. While different disciplines can enrich knowledge about corruption, it may be inadequate to understand the dynamic of this issue in Africa. There seems to be an overreliance on positivism to view corruption problems in higher education. This has led to researchers investigating the existence of corruption in Africa by using international indicators to measure Africans' views on corruption in their society. This section argues for a holistic view of researching the complex reality of corruption in Africa.

Transdisciplinarity in researching corruption in African higher education

The growth of numerous scientific methodologies in the natural sciences allowed the transdisciplinary approach to comprehend many issues in higher education settings (Hadorn et al., 2008). According to Roux, Nel, Cundill, O'farrell, and Fabricius (2017), transdisciplinary research in the natural and social sciences enables the systematic examination of knowledge using several disciplines. Simply said, transdisciplinary research could aid in objectively creating applicable knowledge (Schikowitz, 2020). Currently, this approach is being employed in studies on higher education (Tejedor, Segalàs, & Rosas-Casals, 2018). Transdisciplinarity in higher education research uses methods to pinpoint problems while producing knowledge by analysing various sources. Besides, this approach is increasingly preferred for dealing with many social science problems and offers the chance to validate knowledge using other knowledge domains. As a result, the easiest way to demonstrate the validity of African knowledge and higher education research is through disciplinarity design. Disciplinarity in studying corruption in higher education could be utilised to prove African knowledge about complex issues faced by the systems.

It is important to demonstrate the reality of corruption in Africa using transdisciplinarity. How do we develop knowledge that adheres to the moral standards in Africa, where corruption has all but pervaded politics and the way of life? Corruption is the reason for Africa's slow progress. These bodies of knowledge demonstrate their expertise in economics, religion, sociology, history, and higher education administration in Africa (Collins, 2019). The knowledge generated could enable various disciplines to prove that corruption in African higher education is multidisciplinary and can be understood through ethics and moral systems (Pashby & de Oliveira Andreotti, 2016). It is deeply problematic to prescribe ways of managing higher education resources in Africa without first analysing how history and colonisation have influenced the reality of many of these institutions. However, research should also showcase that corruption and the mismanagement of funds are not only peculiar to the reality of Africa. It remains to be determined whether corruption in Africa is more prevalent than elsewhere in the world.

Moreover, it is essential to discover whether the issue is so complex that it needs Western knowledge on managing corruption to reach high moral values in the African higher education system. Could African universities have been excluded from research projects because they are corrupt and incapable of managing resources? This is similar to what Persson, Rothstein, and Teorell (2010) termed a theoretical mischaracterisation of the problem in the Kenyan and Ugandan context. A significant issue in this debate is that corruption seems to be framed as the main logo of Africa. This state of affairs generates a plethora of questions: who has the highest moral code? What qualifies one to teach others about corruption? Is morality in Africa different from the rest of the world? When do we consider corruption bad, and how do we conclude it can affect all other aspects of African society?

Research on corruption

Table 6.1 examines knowledge trends on corruption to understand different views, methods and ways of finding the truth. Corruption forms part of research in higher education settings because financing is problematic in Africa. Unearthing the truth about corruption requires factoring in this issue's objective or subjective nature. In other words, corruption and higher education are not the only identifiable problems in Africa. Much of the research on financing and corruption in African higher education provides a subjective view of the reality of the situation (Osipian, 2007). How Africans have been studied and portrayed in scholarly publications also depicts the subjective views held by Western epistemology on different topics and issues concerning Africa.

Methodology for studying corruption

A standard method used to study the corruption issue in higher education is the quantitative research design (see Table 6.1). Three main patterns emerge from the historical development of quantitative research: statistical processes, measurement, and test procedures. Quantitative analysis became prevalent in statistics in the nineteenth century due to the desire to compare, relate and correlate two or more ideas (Muijs, 2010). Years later, in the early twentieth century, researchers considered comparing the average group scores of different individuals in educational settings. This process entails testing the relationship between two data sets to generate knowledge, which requires the researcher to formulate a hypothesis about this relationship. Subsequently, this hypothesis is either accepted or rejected by the results generated by the data comparison. Thus, relating and comparing ideas between and among groups produced what is known today as general statistics and is being applied to solve statistical research problems in higher education. The method grew into more advanced designs, leading to different ways of testing variables and relationships (Ary, Jacobs, Irvine, & Walker, 2018). For most research on corruption in Africa (see Table 6.1), researchers usually used a poll primarily focussing on Western-developed measures to demonstrate the extent of corruption. Although African corruption cases have been proven using the quantitative approach, African science cannot support this method.

Table 6.1 Research on corruption in African higher education

Location	*Source*	*Knowledge addressed*	*Knowledge analysis*	*Method of knowing*	*Citation*
International	Report	Academic integrity and dishonesty.	Corruption in higher education is a global problem, but institutions in developing countries face unique challenges. Examples of corruption were more for Africa.	Historical data such as statistics were used to show that corruption is a significant problem in higher education in Africa.	Kirya (2019)
International	Report	Corruption in Africa	Recognises the universality of knowledge of corruption as not limited to politics and geographic locations, yet it focuses on Africa as a challenge.	Conference proceeding on different views about corruption in African higher education supported by evidence such as secondary quantitative data.	Nduku & Makinda (2014)
Nigeria	Journal	Corruption in the management of higher education funds	Corruption in allocating and utilisation of funds affects the quality of higher education in Nigeria.	Quantitative method was used to test the hypotheses and obtain the results about the funding management and declining quality of Nigerian higher education in Nigeria.	Ololube (2016)
South Africa	Journal	Post-apartheid corruption	Corruption at the political level and reasons for student protests, including low funding for higher education.	Political economy and social movement theory were used to study the reason for student protests.	Badat (2016)
Kenya	Journal	Quality affected by resources management and governance	Corruption is coined as mismanagement of resources and ultimately affects governance of higher education in Kenya.	Policy review and analysis of government documents, including statistical information.	McCowan (2018)
Egypt	Journal	Broader topic of corruption in different sectors in Egypt	Corruption caused most rallies in Egypt, and education was also affected.	International indicator of measuring corruption used.	Fayed (2017)

Quantitative design versus historical method

The quantitative method, a unique mathematical approach to solving problems, also involves the application of mathematical explanations of quantifications, which allow researchers to arrive at systematic and substantiated judgments about issues (Zikmund, Babin, Carr, & Griffin, 2013). This provides researchers with a wide application of quantitative methods for evaluation and analysis of corruption in Africa. Using quantitative methods to assess corruption is exciting but may not be adequate in Africa. In essence, the need to evaluate whether the African continent as a whole is corrupt by sampling the opinion of others based on specific indicators is inadequate to prove knowledge of corruption in Africa. The logic of using the quantitative method alone to study corruption in Africa, which has a vast history, cannot be validated by only counting the opinion of people who believe Africa is corrupt without understanding the history of Africans. The reasoning behind uncovering the truth by quantitative means lies in the likelihood of African higher education being corrupt and mismanaged due to the continent's underdevelopment. While there might be some truth to this logical thinking, this positivist thinking and method of deducing the truth cannot produce holistic views of the corruption issue and resource mismanagement in African higher education.

As a result, a southern theory and practice of thoroughly analysing this topic could be based on historical methods or storytelling about the effects of colonialism and how it weakened Africa and its ideals. In terms of knowledge production in higher education, the acceptance of Western notions of studying corruption as the science of conducting research is a power issue. The terms corruption and poverty are defined by Western knowledge and indicators in such a way that excludes their own types of poverty and corruption. Any attempt to disprove this Western logic invalidates and nullifies non-Western science because such a science is not an accepted universal approach. Simply put, alternative epistemological access, analysis, and interpretation are required for African higher education's science of knowledge. The claim that science seeks truth and that this truth is universal has many flaws. To highlight the issues from the ground up, an effort should be made to communicate the science and the truth about financing and corruption in higher education more through stories, history, and occasionally using archaeological evidence found in Africa.

The history of knowledge and the historical development of colonialism in Africa are the reason for corruption in the higher education systems. Despite Africa's numerous resources, colonisers endeavoured to prevent the continent's development. Instead, colonists fought over Africa's resources and occupied innumerable lands. Corruption was first introduced by annihilating African culture, language, and resources, which are other objectives of land conquest. The source of corruption in Africa lies in the West as it used social manipulation to take advantage of Africans and their society. Some Africans who have witnessed thefts and manipulation of the West in changing African culture, history, and knowledge view the period of fighting and conquest for resources as a period of learning. This manipulation was achieved through religion and schooling systems that introduced Western culture and values as true morality and valid knowledge.

The argument is that the research designs used in the study of corruption should primarily start from or be based on the history of Africa. Meanwhile, recent years have seen the development of more complex and challenging approaches. What's more, the late nineteenth century saw the advent of various investigations into societal problems and higher education. A century later, precisely at the beginning of the twentieth century, researchers began to apply knowledge to research complex problems in higher education. In the twentieth century, the knowledge of experimenting and comparing different issues, such as whether corruption in Africa is genetically predetermined, gained momentum. The various developmental stages in experimental designs brought about the use of quantitative design in higher education today. In contemporary times, numerous quantitative studies are being promoted to study most social problems, including in Africa. This is deeply problematic. Qualitative designs increasingly feature in research studies as the dominant method of acquiring knowledge about most complex issues. Historical records of qualitative research showcase fundamental differences between the types of knowledge that cannot be produced from counting knowledge. Understanding these differences requires classifying each research type's main characteristics and attributing these characteristics to research processes in higher education. This classification will facilitate a deeper exploration of the corruption issue. The importance of such a process lies in allowing higher education research to have different methodologies to arrive at the truth about corruption.

Conclusion

The chapter problematises corruption in higher education from a philosophical stance and discusses history's preponderant role in better understanding its knowledge. The analysis shows that the African foundation of knowledge was altered and corrupted based on history. For instance, Africans were provided with an education solely crafted for Western development. Africans were exposed to value systems that promoted the need for the West's stamp of approval on African ways of knowing. As a result, ideologies contrary to Western Science were penalised. The resources used to develop the West also controlled Africans that attempted to challenge it. These mentalities are carried forward into the post-colonial era of African higher education systems, where Africans unconsciously continue formulating policies for Western development.

There are several problems related to resources and corruption in African higher education. One is the [proper] use of resources to achieve a quality that reflects a society's development. Regardless of higher education managers' educational backgrounds and experiences, their minds seem to be focused on propagating Western beliefs and value systems. For instance, the allocation of resources for research in Africa is inadequate, physical infrastructure is not suitable for teaching and learning, and funding for development continues to deplete due to corruption.

The argument is that corruption is universal. However, corruption in Africa has roots in the colonial era but can be undone through engagement with African truth.

Therefore, decolonising minds before tackling corruption of any kind in Africa remains valid. Decolonising the minds to address corruption in African higher education is recognising the truth and reality of African history in terms of knowledge as fundamental to transformation. This requires a grassroots approach where African epistemology, culture and values become ways of life.

Meanwhile, the colonial higher education curriculum is to blame for value systems making Africans concerned with wealth accumulation. The goal of postcolonial higher education is to improve students' socioeconomic standing. Additionally, character and integrity are no longer taught in higher education, which is essential for combating corruption. Assuming that most African leaders are dishonest and that many hold degrees from both African and foreign universities, corruption is a psychological issue affecting everyone concerned. A pressing query is how higher education might decolonise its curricula to promote mental reorientation and dispel myths that Africa is inherently corrupt.

References

Adejumobi, S. (2017). Knowledge for sale? The politics of university education reform in Africa, with a Nigerian example. In *Manoeuvring in an environment of uncertainty* (pp. 179–206). London, England: Routledge.

Agbo, S. A. (2005). Myths and realities of higher education as a vehicle for nation building in developing countries: The culture of the university and the new African diaspora. In *Contemporary issues in education* (pp. 49–69). Leiden, The Netherlands: Brill.

Aina, T. A. (2010). Beyond reforms: The politics of higher education transformation in Africa. *African Studies Review*, *53*(1), 21–40.

Allais, L. (2016). Problematising Western philosophy as one part of Africanising the curriculum. *South African Journal of Philosophy = Suid-Afrikaanse Tydskrif Vir Wysbegeerte*, *35*(4), 537–545.

Ary, D., Jacobs, L. C., Irvine, C. K. S., & Walker, D. (2018). Introduction to research in education. Cengage Learning:Boston, Massachusetts

Aziz, J., & Hashim, F. (2022). Exploring global decolonising projects and English studies in the 21st century: A thematic analysis. *GEMA Online Journal of Language Studies*, *22*(2), 128–140.

Badat, S. (2016). Deciphering the meanings and explaining the South African higher education student protests of 2015–16. *Pax Academica*, *1*(1), 71–106.

Broudy, H. S. (2017). Types of knowledge and purposes of education. In *Schooling and the acquisition of knowledge* (pp. 1–17). London, England: Routledge.

Brunsdon, A. R. (2019). 'Selfishly backward' or 'selflessly forward?': A white male's insider perspective on a challenge and opportunity of decolonisation for practical theology in the South African context. *HTS Teologiese Studies/Theological Studies*, *75*(2), 1–8.

Buba, M. (2020). Mutuntaka 'Personhood' as a paradigm of African epistemology and its implication as a framework of governance. *Asian Journal of African Studies*, (47), 145–170.

Callewaert, S. (2019). Which way Namibia?: Or how to decolonize the colonised mind of the anti-colonial teacher? 1. In *Democratic teacher education reform in Africa* (pp. 222–247). New York: Routledge.

Chimakonam, J., & Ogbonnaya, L. (2021). *African Metaphysics, epistemology and a new logic*. Cham, Switzerland AG: Springer.

Collins, R. (2019). *The credential society: An historical sociology of education and stratification*. New York: Columbia University Press.

Connell, R. (2020). *Southern theory: The global dynamics of knowledge in social science*. London, England: Routledge.

Cook, S. C. C. (2021). *The knowledges that first-year art and design students bring to their university studies*. University of Johannesburg (South Africa). ProQuest Dissertations Publishing, 28935565.

dos Santos Soares, M. A. (2019). On the colonial past of anthropology: Teaching race and coloniality in the global South. *Humanities*, *8*(2), 88.

Drame, I., Connor, S., Hong, L., Bimpe, I., Augusto, J., Yoko-Uzomah, J. … Tofade, T. (2019). Cultural sensitivity and global pharmacy engagement in Africa. *American Journal of Pharmaceutical Education*, *83*(4), 7222.

Erasmus, M. N. (2020). Western ascendency and African capitulation. *Journal of African Foreign Affairs*, *7*(2), 21–33.

Fagbayibo, B. (2019). Some thoughts on centring Pan-African epistemic in the teaching of public international law in African universities. *International Community Law Review*, *21*(2), 170–189.

Fayed, A. A. (2017). The current status of corruption in Egypt. *Contemporary Arab Affairs*, *10*(4), 510–521.

Hadorn, G. H., Biber-Klemm, S., Grossenbacher-Mansuy, W., Hoffmann-Riem, H., Joye, D., Pohl, C. … Zemp, E. (2008). The emergence of transdisciplinarity as a form of research. In *Handbook of transdisciplinary research* (pp. 19–39). Dordrecht: Springer.

Harnois, Y., & Gagnon, S. (2022). Corruption and international development: A review of project management challenges. *Journal of Financial Crime*, *29*(3), 864–877.

Heleta, S. (2016). Decolonisation of higher education: Dismantling epistemic violence and eurocentrism in South Africa. *Transformation in Higher Education*, *1*(1), 1–8.

Higgs, P. (2016). The African renaissance and the transformation of the higher education curriculum in South Africa. *Africa Education Review*, *13*(1), 87–101.

Hwami, M. (2023). Authoritarian neoliberalism, kleptocracy and the complicity of the African academe. Amasa P. Ndofirepi, Felix Maringe, Simon Vurayai, Gloria Erima (Eds.). In *Decolonising African university knowledges, Volume 2* (pp. 51–69). Routledge: New York.

Joseph, C. L. (2018). Toward a black African theological anthropology and Ubuntu ethics. *Journal of Religion and Theology*, *2*, 16–30.

Kirya, M. (2019). Corruption in universities: Paths to integrity in the higher education subsector. U4 Anti-corruption resource centre. U4 Issue (10). https://www.u4.no/publications/corruption-in-universities-paths-to-integrity-in-the-higher-education-subsector

Knaus, C. B., Mino, T., & Seroto, J. (Eds.). (2022). *Decolonising African higher education: Practitioner perspectives from across the continent*. London, England: Taylor & Francis.

Laga Ramoupi, N. L., & Ntongwe, R. N. (2017). In M. Cross & A. Ndofirepi (Eds.), Africanisation of humanities knowledge in the universities in Africa: A critique of the Cameroon and South African experiences. In *Knowledge and change in African universities* (pp. 195–214). Leiden, The Netherlands: Brill.

Lassou, P. J., Hopper, T., Tsamenyi, M., & Murinde, V. (2019). Varieties of neo-colonialism: Government accounting reforms in anglophone and francophone Africa–Benin and Ghana compared. *Critical Perspectives on Accounting*, *65*, 102071.

Luyaluka, K. L. (2016). An essay on naturalized epistemology of African indigenous knowledge. *Journal of Black Studies*, *47*(6), 497–523.

Maistry, S. M. (2021). Threats to decolonial immanent ethics. *African Journal of Inter/Multidisciplinary Studies*, *3*(1), 26–37.

McCowan, T. (2018). Quality of higher education in Kenya: Addressing the conundrum. *International Journal of Educational Development*, *60*, 128–137.

Mede, N. G., & Schäfer, M. S. (2020). Science-related populism: Conceptualizing populist demands toward science. *Public Understanding of Science*, *29*(5), 473–491.

Mittelman, J. H. (2017). Decolonizing the postcolonial university? Possibilities and exigencies with evidence from Uganda. *Journal of Poverty Alleviation & International Development*, *8*(1), 1–47.

Mohamedbhai, G. (2020). The scourge of fraud and corruption in higher education. In *Corruption in higher education* (pp. 138–143). Leiden, The Netherlands: Brill.

Mudimbe, V. Y. (2020). *The invention of Africa: Gnosis, philosophy, and the order of knowledge*. Lulu Press, Inc: Morrisville, North Carolina, USA.

Muijs, D., Ainscow, M., Chapman, C., & West, M. (2011). Collaboration and networking in education. London: Springer.

Mzwazi, J., & Shumba, S. (2022). Religiosity and African epistemology. In D. Masaka (Eds.), *Knowledge production and the search for epistemic liberation in Africa* (pp. 149–165). Cham, Switzerland: Springer.

National Academies of Sciences, Engineering, and Medicine. (2018). In D. Skorton & A. Bear (Eds.),The integration of the humanities and arts with sciences, engineering, and medicine in higher education: Branches from the same tree.

Ndelu, S. (2017). A rebellion of the poor: Fallism at the Cape Peninsula University of Technology. *Hashtag: An analysis of the# FeesMustFall movement at South African universifies*, 13–32.

Ndlovu-Gatsheni, S. J. (2015). Decoloniality as the future of Africa. *History Compass*, *13*(10), 485–496.

Ndlovu-Gatsheni, S. J. (2018). The dynamics of epistemological decolonisation in the 21st century: Towards epistemic freedom. *The Strategic Review for Southern Africa*, *40*(1), 16–45.

Ndofirepi, A. P. (2023). Questioning the neoliberal agenda: What is to know in the African university? In A.P. Ndofirepi, F. Maringe, S. Vurayai, & G. Erima (Eds.), *Decolonising African university knowledges, volume 2* (pp. 1–10). London, England: Routledge.

Ndofirepi, A. P., Mathebula, T., & Gwaravanda, E. T. (2020). On relevance: A critique of knowledge processes in the African university. In A. P. Ndofirepi & M. Musengi (Eds.), *Inclusion as social justice* (pp. 71–89). Leiden, The Netherlands: Brill.

Nduku, E., & Makinda, H. (2014). Combating corruption in society: A challenge to higher education in Africa. Corruption in Africa, 281. http://iffoadatabase.trustafrica.org/iff/Corruption_in_Africa_A_Threat_to_Justice_and_Sustainable_Peace.pdf#page=286

Nell, I. A. (2021). Decolonising an introductory course in practical theology and missiology: Some tentative reflections on shifting identities. *Transformation in Higher Education*, *6*, 103.

Nkulu-N'Sengha, M. (2005). African Epistemology. *Encyclopedia of Black Studies*, *39*, 44.

Nyamnjoh, F. B. (2019). Decolonizing the university in Africa. Thompson, William R (Ed.). In *Oxford research encyclopedia of politics*, 164–171. Oxford, England: Oxford University Press.

Odaga, G. (2019). *Achieving equity and gender equality in Uganda's tertiary education and development.* (Doctoral dissertation). University of South Africa, Pretoria.

Ololube, N. P. (2016). Education fund misappropriation and mismanagement and the provision of quality higher education in Nigeria. *International Journal of Scientific Research in Education*, *9*(4), 333–349.

O'Neal, S. (2020). Pan-African aesthetic: Pan-Africanism in afro-beat. In R. Rabaka (Ed)., *Routledge handbook of pan-Africanism* (pp. 489–502). London, England: Routledge.

Osipian, A. L. (2007). Corruption in higher education: Conceptual approaches and measurement techniques. *Research in Comparative and International Education, 2*(4), 313–332.

Pashby, K., & de Oliveira Andreotti, V. (2016). Ethical internationalization in higher education: Interfaces with international development and sustainability. *Environmental Education Research, 22*(6), 771–787.

Persson, A., Rothstein, B., & Teorell, J. (2010). The failure of anti-corruption policies: a theoretical mischaracterization of the problem. QoG working paper series 2010:19, University of Gothenburg, Gothenburg, Sweden.

Roux, D. J., Nel, J. L., Cundill, G., O'farrell, P., & Fabricius, C. (2017). Transdisciplinary research for systemic change: Who to learn with, what to learn about and how to learn. *Sustainability Science, 12*(5), 711–726.

Salehyan, I., Hendrix, C. S., Hamner, J., Case, C., Linebarger, C., Stull, E. … Williams, J. (2012). Social conflict in Africa: A new database. *International Interactions, 38*(4), 503–511.

Sanneh, L. (2015). *West African Christianity: The religious impact*. Maryknoll, NY: Orbis Books.

Schikowitz, A. (2020). Creating relevant knowledge in transdisciplinary research projects-coping with inherent tensions. *Journal of Responsible Innovation, 7*(2), 217–237.

Sklar, R. L. (2015). *Nigerian Political parties: Power in an emergent African nation* (Vol. 2288). Princeton, NJ: Princeton University Press.

Smith, C. (2014). *Disruptive religion: The force of faith in social movement activism*. New York: Routledge.

Spivak, G. C. (2016). Humanities, democracy and the politics of knowledge in higher education. Michael Anthony Samuel, Rubby Dhunpath, Nyna Amin (Eds.). In *Disrupting higher education curriculum* (pp. 17–30). The Netherlands: Brill.

Tejedor, G., Segalàs, J., & Rosas-Casals, M. (2018). Transdisciplinarity in higher education for sustainability: How discourses are approached in engineering education. *Journal of Cleaner Production, 175*, 29–37.

Tella, O. (2021). South Africa's curriculum transformation: Insights from post-independence Africa and post-civil rights movement in the United States. *Strategic Review for Southern Africa, 43*(2), 242–266.

Terblanché-Greeff, A.C. (2019). Ubuntu and environmental ethics: The West can learn from Africa when faced with climate change. In M. Chemhuru (Ed.), *African environmental ethics. The international library of environmental, agricultural and food ethics* (Vol. 29). Cham, Switzerland: Springer. DOI: 10.1007/978-3-030-18807-8_7.

Unterhalter, E., Allais, S., Howell, C., McCowan, T., Morley, L., Ibrahim, O. … Oketch, M. (2018, March). Conceptualising higher education and the public good in Ghana, Kenya, Nigeria, and South Africa. In *Proceedings of the CIES 2018 annual conference* (Vol. 2018). Mexico City, Mexico: Comparative and International Education Society (CIES). Teacher Education, Columbia University.

Uslaner, E. M. (2017). *The historical roots of corruption: Mass education, economic inequality, and state capacity*. Cambridge, England: Cambridge University Press.

Young, C. (2004). The end of the post-colonial state in Africa? Reflections on changing African political dynamics. *African Affairs, 103*(410), 23–49.

Zikmund, B. Carr & Griffin (2013) Business Research Methods. Fisher, M.(9th ed.), Cengage Learning.: SOUTH-WERTERN: Canada.

Index

For Product Safety Concerns and Information please contact our EU
representative GPSR@taylorandfrancis.com
Taylor & Francis Verlag GmbH, Kaufingerstraße 24, 80331 München, Germany

www.ingramcontent.com/pod-product-compliance
Lightning Source LLC
LaVergne TN
LVHW021247011026
849160LV00010BA/977
9781032428840